Tea on the terrace

Manchester University Press

Tea on the terrace

Hotels and Egyptologists' social networks, 1885–1925

Kathleen L. Sheppard

Manchester University Press

Published by Manchester University Press
Oxford Road, Manchester M13 9PL

www.manchesteruniversitypress.co.uk

British Library Cataloguing-in-Publication Data
A catalogue record for this book is available from the British Library

ISBN 978 1 5261 6220 3 hardback
ISBN 978 1 5261 7889 3 paperback

First published 2022

Typeset
by New Best-set Typesetters Ltd

Contents

Figures

Tables

Acknowledgements

I have been thinking about the ideas in this book since 2010, and, obviously, not on my own. There are so many people to whom I owe gratitude; I hope this partial list will begin to thank them for all they have done. Also, as my own research has shown me time and again, this section of every book is useful in establishing networks of contact that would not otherwise be present in any other document.

In the end, little of this book is based on readily available published works. I have depended on the kind and enthusiastic expertise of archivists all over the world. The staff at the Egypt Exploration Society, especially former director Chris Naunton, current director and former archivist Carl Graves, and current archivist Stephanie Boonstra have sought out hundreds of letters and shared their deep and wide knowledge about everything EES and EEF. Anne Flannery and John Larson of the archives at Oriental Institute at the University of Chicago have shared secrets the archives don't always reveal and for that I am thankful. Paul Smith, former archivist at the former Thomas Cook archives, really should write his own book; I counted on the lunches we had in Peterborough for guidance through the timeline I was working through and, of course, the things only a good archivist knows about their material. So many thanks to the Griffith Institute, especially Cisco Bosch-Puche and Liz Fleming, who kept talking about my project when I would pop out for a snack and have new things for me to see when I would come back. The Egyptian Art department at the Metropolitan Museum of Art welcomed me into their archive and let me look at so many treasures; Diana Patch and Catharine Roehrig kindly allowed permission to quote from the collections. The staff at the Garstang Museum, the Brooklyn Museum, the New York Historical Society, the British Library, and the British Museum freely gave their materials so I could find (most of the time) what I was looking for; sometimes, materials just didn't make it into the archive. Finally, the History of Science collections at my alma mater, the University of Oklahoma, and their staff, Cassondra Darling, Kerry Magruder, Melissa Rickman, and JoAnn Palmeri, hosted me many times over the last decade and this project is all the richer for the knowledge networks I was a part of there. Thank you to all of these archives, libraries, and collections for letting me quote from their materials.

I have presented these ideas in various forms at conferences and annual meetings, including the Midwest Junto for the History of Science, the History of Science Society, the European Association of Archaeologists (EAA), the American Research Center in Egypt (ARCE), the Histories of Archaeology Research Network (HARN), the Association for the Study of Egypt and the Near East (ASTENE), the Humanities and Technology Conference, and the Missouri Egyptology Symposium. I was also able to speak to public audiences at the Linda Hall Library in Kansas City, the Egypt Study Society in Denver, Colorado, Colorado School of Mines in Golden, Colorado, and the Friends of the Petrie Museum in London. I have received immeasurably helpful and encouraging feedback from all of these groups and I thank you for your time and thoughtfulness. Andrew Humphreys gave me the initial encouragement and guidance to people and archives I didn't know I needed. Sarah Ketchley offered insight into life on the *Beduin* through her encyclopaedic knowledge of Andrews' diaries. Ziad Morsy's work on Nile boats and the conversations we had were central to the third chapter, especially. Will Carruthers' and Gabe Moshenska's research, along with their support over the years, has been invaluable.

My resources were spread over the US, the UK, and Egypt, so funding was absolutely necessary. In 2018, I received generous funding from the Andrew Mellon Foundation at the University of Oklahoma to spend two weeks in the collections there; I also presented to my colleagues there, and their feedback was crucial to the way I thought about these issues. Margaret Gaida left me many helpful voicemails; Kathleen Crowther told me to keep following the circles I was stuck on, and I am so glad she did. I was able to spend a week in Oxford as a Harris-Manchester Fellow in 2019. My own institution has provided support in the form of internal university research grants, and in 2021 the Christensen Fellowship allowed me to tie up a few loose ends. In each of these places I met and talked with people who made this project better, and the sometimes lonely research process richer for their company, especially Ann Keller.

Thank you, friends and colleagues who have listened to me talk about this project for so many years; a special thank you to the ones who braved a draft reading at some point. Thank you to my Lady Science crew: Robert Davis, Samantha Muka, and Joy Rankin, who helped me clearly articulate what I wanted to say in my introduction; and as always to Rebecca Ortenberg, Leila McNeill, and Anna Reser for allowing me to work through some of my ideas in a short piece or two. Katherine Pandora read and commented on many parts of this book; her comments, in that familiar black ink, have marked almost everything I've written since 2004 and I cannot thank her enough. Jeanine Bruening's amazing questions on the introduction and first chapter helped me to reframe many of the questions I was asking. Michael Robinson didn't just read chapter drafts, he read a proposal draft and that's a special kind of friend. Ben Gross, thank you for answering the phone one day when I called randomly and for your encouragement to continue working on this project, because it really did matter. Alanna Krolikowski hasn't read these words, but she has sat through a lot of me telling her about this and I thank her for her patience and good advice. Julia Roberts, you read every single word and your frank thoughts, comments, and questions are throughout these pages. Thank you so much for answering my text messages. Thank you also to the anonymous reviewers whose suggestions enriched the narrative. Even with all of this amazing input, my mistakes are my own.

Thank you to the people who have sustained me through this, both the work and the life that goes with it. My Friday pandemic writing group who saw me through a lot of these chapters: Elaine Ayers, Eddie Guimont, Alison Laurence, Ada Link, Lydia Pyne, and Anna Toledano know how all of this has gone. Sarah Naramore, Sarah Pickman, and Sarah Qidwai have given support, friendship, and encouragement from a safe, social distance. Dan Reardon is almost more excited about this than I am. Marikay Asberry, Amanda Byrne, Summer Chapman, Beth Dare, KC Dolan, Kristen Gallagher, and Sam Welter for the daily sustaining support. Anne DeWitt has offered partnership on another project and friendship in real life. Support from the ARCE-MO chapter – Anne Austin, Stacy Davidson, Lisa Haney, Rozanne Klinzing, Julia Troche, and Clara Wright – has been immeasurable.

Personally, I need to thank so many people for so many things. They appear a lot in my acknowledgements but life and this book would be worse without them: Cornelia Lambert, Lynnette Regouby, Sylwester Ratowt, and John Stewart are my grad school cohort whose support and understanding ground me. Meegan Neeb travelled with me to Egypt in 2015 to wander around these spaces with me – let's do it again! Mary Hughes, Jeff Langman, and Usha Natarajan, my partners in crime during our Cairo days, were there during multiple flat tyres on the Ring Road and so much falafel. Tom Aldred and Claire Thornhill, and later their two boys, made London home for me by making their home mine for the last twenty years. Hannah Cordts and her boys feed me yummy dinners when I'm in London. Running with Pennie Vargas all over London to come back to breakfast with Rhiannon Lloyd and their family is tradition. Thank you Chris Naunton for letting me bully you into a podcast over the pandemic, and for meeting up with me in London over the years for drinks and snacks and a nice walk to Highbury. Gemma Renshaw, thank you for letting me into the Petrie Museum to see wonderful things (no, not those) and your *Assassins' Creed* skills.

Thanks always to Courtney, Amy, Olivia, and Will Brackin, my Oklahoma family, for always letting me stay with you and for making me a part of your family. Ron and Linda Harris have done a lot of childcare and pet care over the years and we miss you being so close. Kate Drowne, Shannon Fogg, Audra Merfeld-Langston, and Kris Swenson who are my trusted mentors, writing accountability partners, and inappropriate friends. Julie Jonsson, who listens to my voice messages all the time and now that we're far away I love hearing your voice on my phone. My dad passed away a few years ago, when I was really in the thick of writing. His love and lifetime of support for whatever crazy thing I wanted to do next are woven into everything I do. Mel Holbrook, thank you for continuing that level of support; I'm so thankful for you.

With a book that has been in progress so long, the acknowledgements are going to be long. But they wouldn't be complete without thanking my husband and son. They are both patient and kind when I have my Bodleian Library 'silence please' sign on my door. They don't always abide by it, but they understand why it's there. I have a wall full of hearts and love notes from my son, who is now seven years old, and whose own school desk was in my office with me from March to June of 2020, and a few other times after that. I couldn't ask for a better office mate. I love you infinity thousand.

Abbreviations and a note on spelling

Andrews' Diary	*A Journal on the Bedawin 1889–1912* by Mrs Emma Andrews, courtesy of The Metropolitan Museum of Art, Department of Egyptian Art
Baedeker's *Egypt*	Karl Baedeker, ed. *Egypt: Handbook for Travellers*. Leipsic: Karl Baedeker. Years vary
Murray's *Handbook*	*A Handbook for Travellers in Lower and Upper Egypt*. London: John Murray. Years vary

A note on spelling

I have kept the original spellings of place names as written by the original authors, but I have tried to present the current transliterations in my text.

Currency conversion table

Table 0.1 Currency conversion, by year, based on Baedeker's *Egypt* editions

Edition year (pages)	Egyptian Pounds (Livre Egyptienne, £E)[a]	Great British Pounds (GBP, £)	United States Dollars (USD, $)	French Franc[b]	GBP today (2019)[c]	USD today (2021)[d]
1878 (3–5)	195	1	4	25.25	70.14	105.95
1885 (3–5)	195	1	4	25.25	70.14	109.23
1895[e] (i; xvi)	1	20s. 6d.[f]	5	25.95	89.13	157.66
1898 (i; xviii-xix)	1	20s. 6d.	5	25.95	84.92	159.56
1902 (i; xv)	1	20s. 6d.	5	25.90	85.35	154.00
1908 (i; xv)	1	20s. 6d.	5	25.90	84.92	143.95
1914 (i; xv)	1	20s. 6 ¼d.	5	25.90	64.09	132.44

[a] In 1878 and 1885, the Piastre was the main currency, with each Piastre containing 40 paras. See note e, below, for currency change.
[b] French francs were the common currency for the Latin Monetary Union.
[c] Purchasing power found at the National Archives Currency Converter: 1270–2017, www.nationalarchives.gov.uk/currency-converter/ (accessed 22 April 2021); brought up to most recent data (2019) at Measuring Worth: www.measuringworth.com/calculators/exchange/ (accessed 22 April 2021).
[d] Purchasing power found at the CPI Inflation Calculator, www.in2013dollars.com/ (accessed 22 April 2021); compared to GBP at Measuring Worth: www.measuringworth.com/calculators/exchange/ (accessed 22 April 2021).
[e] In 1895, Baedeker notes: 'The confusion that formerly reigned in the Egyptian currency has been removed by the introduction of a new national coinage' (xvi).
[f] This is equal to £1 (GBP).

Introduction: archaeologists in Egypt

On his first trip to Egypt in 1880, Flinders Petrie saved time and money by staying in the rock-cut tombs at Giza. He wrote that 'no better lodgings are to be had anywhere for solidity and equable temperature; the minor advantages may be a question of taste, such as the gratis supply of ancient bones or mummy cloth in the dust and sand of your floor'.[1] He knew there were other options, but, he told his readers, for the money, 'the tomb is best'.[2] There, he was free of interruptions from company, except for the occasional tourist or official visitor, his small crew, and the goat he kept for fresh milk. Petrie would become infamous for his frugality and distaste for the comforts of urban life.

Unlike Petrie, most Egyptologists and archaeologists arriving in Egypt in the late nineteenth and early twentieth centuries chose to stay in hotels when they arrived in the urban areas of Alexandria, Cairo, and Luxor in the days and weeks before their season began. Before they could begin excavations, they needed to gather supplies, organise transportation, hire and prepare work crews, and obtain excavation permissions. In preparation for the season's field work, many archaeologists also spent time working in the Cairo Museum or purchasing artefacts from dealers. These tasks were essential to archaeologists preparing for a long and isolated field season, which lasted through the cooler months, from October to April, in Egypt. Depending on where their work was located, archaeologists might also stay in hotels while excavating. This was especially true for those working in and around Luxor, where work sites might be within reasonable walking distance of a hotel.

During their preparation time, these archaeologists formed small, ephemeral communities of scholars who frequently met for parties, teas, drinks, meals, and – most importantly – conversation. They shared news from home, gossip about colleagues and friends, ideas about archaeology, theories about excavation sites, and plans for their work. These discussions occurred far from the constrictions and conventions that governed the archaeologists' lives and work back in their home institutions. These activities not only reflected the ever-changing environment in which they took place, but they also shaped the way in which science was conducted in that relatively uncontrolled environment, and beyond. As the sites of such activities, Egyptian hotels, I argue, functioned as Egyptological think-tanks. Egyptology began and operated under the umbrella of a European colonial system, and for the time period in this book, specifically British colonial power. In that context, I analyse the influence of ephemeral hotel spaces in the networks formed within them and the interpersonal performances within the places, groups, and networks up and down the Nile.

Historians have shown that European science was a product of place: the breakfast table, the home salon, the pub, and the coffee house.[3] In these spaces, which accommodated both the public and the private, people could meet to chat, to read newspapers and pamphlets, and to discuss philosophy, politics, and science. From the seventeenth to the nineteenth century and beyond, people of all classes and genders had access to the debates in various public meeting places, and women were especially active in those debates that occurred within the home.[4] In the history of archaeology, too, place has power in the production of knowledge. Among archaeologists, the knowledge gained at the trowel's edge has always held particular importance; scientists developed and refined their theories based on objects coming out of the ground. Recently, scholars have highlighted the importance of other locations where archaeological knowledge is created, including the classroom, in which future practitioners are educated in the field, and the museum, in which the general public is exposed to archaeological theory and practice.[5] This book argues that Egyptian hotels represent an important locus for the creation of archaeological knowledge and the networks necessary for that knowledge.

The chapters that follow discuss the hotels themselves as central spaces in the science of archaeology and the activities of their temporary residents as behaviours that drove the development of that science. As they gathered in Egyptian hotels each autumn, they formed a dynamic cognitive topography that they later carried with them into the field.[6] Martin Rudwick's work has addressed the issue of cognitive topographies in the evolutionary sciences of the nineteenth century.[7] He argued that the 'map of the *social* field of the science is at the same time a map of its *cognitive* topography'. It follows, then, that that landscape 'encapsulates in descriptive terms, and to some extent in explanatory terms, the ways in which the treatment of knowledge claims depended on their point of origin'.[8] In Egyptology, these social networks that steered the science for decades had clear and distinct points of origin for their particular knowledge claims – hotels. In these hotels, Egyptologists experimented with ideas and expanded and refined their theories and methods. They formed intellectual networks and banded together as teams for field work. They found patrons, employees, colleagues, friends, and lovers. They administered their excavations from and within the friendly confines of hotel rooms and gardens. They counted on the colonial power of these Western hotels to form a barrier between themselves and Egyptians. Hotel environments offered a space largely unconstrained by expectation for professional behaviour, allowing archaeologists to perform a number of crucial professional tasks without the limitations of institutional expectations.[9] Using the framework of scholarship on geographies of knowledge creation in science and on social and professional networks in science alongside an examination of the professional behaviours of many archaeologists working in Egypt, I will show that the hotel environment should be centralised by historians because it showcases the sociality and sociability of science, illuminating how scientists play with and produce knowledge in the most casual of settings.

A reappraisal of Egyptology

The chapters below present a brief history of Egyptology within a history of tourism in archaeology.[10] This analysis encounters Egypt between the years 1885 and 1925 as a historical geography and phenomenological travelogue of Egypt and its sites through the eyes of some of its European visitors. This book, then, is a history of travel and social networks in Egypt and Egyptology as much as it is an examination of the role of hotels in this particular place and discipline. This book combines the history of travel in Egypt with a study of Egyptology through the framework of the social studies of science. Using this context, I aim to understand the nature of the practice of science itself, which includes social, political, and economic factors. Egyptology is full of opportunities to study explicitly the impact of networks, spaces, and discussion on scientific knowledge.[11] This is not to say that putting the science of Egyptology in its place has never been done; in fact works about Egyptology and colonialism, Orientalism, economics, politics, and social movements abound.[12] There are analyses of how the physical landscape of Egypt affected the discipline, which are crucial to understanding what drew and still draws people to the Nile Valley.[13] Additionally, there is no shortage of fascinating biographies of practitioners and, included in them, beautifully written thick descriptions of tombs, temples, and excavation activities.[14] These histories are essential to piecing together activities and outcomes from particular sites in particular eras. Finally, various institutional histories frame the science of Egyptology using gender, race, class, and socio-political status.[15] These theoretical approaches are useful contributions to our understanding of Egyptian archaeology. Few, however, use the theoretical frameworks of the history and philosophy of science to approach the intricate issues in the discipline.[16]

This book presents a more complete picture of scientific practice than a disciplinary history or biography might. Steven Shapin makes the case that attention to the social context informing science acknowledges the human element in a given discipline and thus offers a more accurate treatment of it. He goes on to say that, when scholars employ new theoretical stances such as history and philosophy of science, it means that they are:

> committed to telling rich, detailed, and, we hope, accurate stories about science without believing that it is cognitively or methodologically or socially unique, without believing that it is integral and unified, without believing that it has a special set of values not possessed by other forms of culture, without believing that it is divinely inspired, without believing that it is produced only by geniuses, without believing that it is the only progressive force in history.[17]

This book, then, attempts to begin a reappraisal of some of the best-known activities of people who worked on major sites whose finds made a significant impact on Egyptology. Many histories of Egyptology mention these hotels, but none centralise them. Some of the people, places, and events I address will be familiar to Egyptologists and devotees of the field, but that is part of the point: I hope to make the familiar new with an analytical shift that privileges the social studies of scientific networks. That is, I hope to demonstrate that *where* knowledge is created has an enormous impact on the *how*, the *what*, the *who*, and the *why* of that knowledge.[18] In addition, this book is similar to other travelogue histories in terms of source material, spaces, and travel trajectory; however, it is different in that it uses a social and historical framework to focus specifically on what the practitioners of Egyptology did in these particularly fraught spaces.

Sites of knowledge creation in science

David Livingstone and others have argued that scientific knowledge is shaped by the location of its making.[19] Therefore, as far as the lab or field scientist was concerned, 'it was *presence* not absence, *closeness*, not distance, that underwrote their claims to authenticity'.[20] Archaeologists making knowledge in the field were clearly imbued with professional status by displaying not only their heroic deeds, but also by producing field reports and new theories nearly every year. As the lab is for the chemist, for the archaeologist, the field occupies the highest and most 'privileged place of knowledge generation'.[21] But to tell the whole story of archaeological science, historians must consider what archaeologists did with the time when they were not at the field site.

Most worked in universities or museums, many of them taking that time to write up their field results. They did not do this in a vacuum; even the most seemingly independent field archaeologists had armies of assistants, curatorial assistants, collections organisers, administrators, translators, and colleagues who formed a community essential to their work. Before presenting their ideas in print form, many gave lectures and attended conferences and scholarly meetings, fine-tuning their theories at length in these professional and controlled settings. However, because knowledge 'bears the imprint of its location', and because the creation of knowledge occurs also at locations outside the halls of academe, it is crucial to take into account locations where archaeologists were not constrained by scholarly convention.[22]

Egypt boasted plenty of coffee and tea houses, but many were exclusive to locals; a foreigner would have been out of place.[23] European and American visitors usually made their way instead to a handful of European-run hotels in urban centres to stay, play, and work. The hotels contained all the elements of the sites where they might have met at home. There were dining rooms for polite discussion over breakfast, tea, and dinner and there were verandas for watching the world go by while drinking coffee or cocktails. All these spaces hosted a wide range of people who offered news, friendship, and sometimes romance, and always the opportunity to debate politics or science.[24] Historian Molly Berger argues that hotels were part of an 'urban cultural landscape' in which it is impossible to extricate the hotel building from the hotel culture.[25] Indeed, in the winter months of the late nineteenth and early twentieth centuries, hotels in Egypt, and especially Cairo, were places to see and be seen for archaeologists as well as tourists. There was a range of comfortable accommodation strictly for Europeans and Americans; these varied from hostels run by priests for 7–8 piastres a day, to luxury hotels like the famous Shepheard's at 50 piastres per day.[26] Andrew Humphreys describes the histories and clientele of many European-run Egyptian hotels from Alexandria in the north to Aswan in the south, demonstrating that these establishments drew not only archaeologists, but also tourists, businessmen, government officials, military personnel, journalists, and antiquarians from all over the world.[27] Many such guests mixed with the archaeologists, participating in and thus shaping scientific activities in important ways; their rich stories can be found in memoirs, private and professional correspondence, and newspaper accounts.

In hotels, as in European coffee houses, scientific activity followed no prescribed formula; archaeologists were unfettered by the formality of the work site or the museum, and they could give their scientific imagination free rein. Further, hotels differed from informal meeting places in Europe in one important respect: they were far from home. As Humphreys noted, the hotels were 'outposts of Europe planted on Egyptian soil'.[28] For archaeologists, being away from the social limitations of their home countries as well as away from the usual professional constraints made hotels the most informal and unrestricted arenas for discussing new theories, establishing new patronage relationships, and creating new scientific networks.[29]

Tea on the terrace: table talk, truth spots, and scientific networks

Interactions between Egyptologists, their excavation crews, and other guests made these hotels truth spots, not unlike universities, museums, or temples.[30] Thomas Gieryn defines truth spots as spaces where ideas 'gain credibility' and that, in turn, function as a 'vital cause of that enhanced believability'.[31] In other words, hotels as truth spots are places 'where an account and a potential believer intersect', making the potential believer more likely to trust certain ideas or assertions simply because of the place itself.[32] In the case of Egyptian hotels, Egyptologists and archaeologists met on the terrace for tea, in ballrooms during dances, in dining rooms for meals, and in their private rooms, all to discuss scientific matters. In doing so, they turned hotels into scientific institutions.[33] They hired and fired excavation crew and made important administrative decisions in these truth spots. Further, in the public spaces in hotels, tourists could encounter information directly from Egyptologists as they created it. Howard Carter's proclamation about closing the tomb of King Tutankhamun was posted first in the Winter Palace Hotel lobby for the purpose of making his decision – and his side of the argument – known to the tourist public.

The people who met in the Egyptian hotels not only formed short-lived communities for discussing science, but also established longer-lasting relationships as well. Rudwick's category of cognitive topographies, in which readers can witness the changing shape of scientific knowledge and networks, is crucial to this discussion as is Michael Farrell's idea that these collaborative circles formed groups in which members came 'to play an informal role in the circle, and … are affected by the group and the roles they play in it'.[34] Beginning with table talk over a meal or tea, the peripheral members would come and go, but the core group of Egyptologists remained constant. It was during these conversations, I argue, that Egyptologists were most free to form these circles, or networks, of collaborators and colleagues.[35] The peripheral members could also integrate themselves within the network, thus eventually becoming central characters themselves.

Once they left the hotels, Egyptologists continued to correspond with each other about their ideas, to seek each other's help, and in many cases, to develop lifelong friendships as well. Because their relationships began informally, they often remained so. As a result, their interactions are not always visible in the published record. Most of the sources I use throughout the book are diaries, correspondence, and other private documents held in small institutional archives. From these sources, I reconstruct the cognitive landscape of particular times, places, and people. Because these are private documents, some characters make appearances where readers might not expect to see them, and others are absent where readers might have presumed to find them. The personal nature of these documents, however, allows me to analyse the relationships within and among these scientific networks as real friendships, and therefore I can attempt to examine the ideas they shared and how those ideas took shape through informal networks over time and distance.

The creation of these networks and communities and the nature of their activities are central to this volume. The history of science has long concerned itself with groups of colleagues, assistants, students, and staff, but in the context of archaeology, the topic demands some explicit discussion. Scholars who study present-day scientific networks argue that the best way to trace connections is through joint publications and reviews of those publications.[36] They argue further that correspondence offers key evidence of how networks interacted behind the scenes.[37] In archaeology, such correspondence is especially important. The most influential participants in archaeological networks tended to gather in the field in transitory groups. Some group members were permanent fixtures every dig season, others came and went, and still others appeared only once, briefly, leaving just the faintest trace in the records. Joint publications do not necessarily reveal the connections among group members, so those connections can be difficult to trace. As Janet Browne has recently argued, studying correspondence among scientific networks offers 'the prospect of reconstructing patterns of sociability with due appreciation to the structure of the society in which they emerged'.[38] These personal communications allow us to gain insight into schools of thought, and thus to understand better who was sharing ideas, how they were being shared, and who was participating.

In Egyptology, knowledge is created, discussed, and refined in both institutional spaces, such as the university or museum office, and in the field. In early Egyptology, network nodes were concentrated in cities all over the world. Societies like the Egypt Exploration Fund (EEF, from 1919 Egypt Exploration Society [EES]), institutions such as the Cairo Museum, the British Museum, the Oriental Institute, the Metropolitan Museum of Art, the Louvre, the Museo Egizio, and the Berlin Museum were all hubs from which and to which scholars and their ideas travelled. As Egyptology's formal institutions, these centres contain records of membership, meeting minutes, details of official activities and decisions, and formal collections of scholarship. The records held by these institutions help to tell a rich story of discipline formation from an official point of view. Their published collections are crucial to historians and other scholars who trace changing ideas over time. Through collected correspondence, historians are also able to view the connections among these metropolitan hubs. It is not only from the content of such letters, but also from the identities of senders and receivers, that readers see who was present and active in multiple networks, what they said to each other, and how they interacted. Many times, the letters reveal that women were actively involved in these institutions, running the administrative side of institutional life while the men were in the field. Women were necessary and important parts of these networks, but they are part of the group that tends to be left out of the story. However, throughout this book I aim to embolden the voices of women such as Emily Paterson of the EES, American patron Emma Andrews, and Margaret Murray of the UCL Egyptology department. Women were instrumental in shaping the discipline of Egyptology, and archival documents underline their activities.

In the field, far from these metropolitan institutions, network developments and movements are harder to trace. But the mundane, everyday activities of these sites were often even more central to the evolution of the discipline than were published scholarship and institutional organisations. Correspondence and field diaries, usually found in archives at the hubs, are crucial tools for tracing network participation in different areas of the scientific practice. Such sources allow scholars to witness the activities of these hub institutions, to see how, when, and where people moved throughout the excavation seasons, and to understand how they used the off-seasons. The evidence of correspondence, field diaries, and other personal documents allows us to trace the creation of scientific knowledge within and among the hubs, as well as in the field, observing how each location produced different evidence, new relationships, and a variety of scholarly outcomes.

The book is both a historical geography and a historical and social analysis of the role of hotels; so, throughout each chapter, I will be sharing archaeologist and tourist experiences in Egypt in the late nineteenth and early twentieth centuries. The views of these visitors were firmly entrenched in the contemporary Orientalist and colonial ideology of Europeans and Americans at the time. Their reactions to Egyptian culture and environment are shocking to the twenty-first-century reader. Scholars, however, have critically examined these views, and it is not the purpose of this book to revisit their work here.[39] This book will, however, place Egyptologists firmly in this context while at the same time continue the process of dismantling and problematising the colonial milieu. This is not a book that makes heroes out of Egyptologists and their activities.

Ever since the first English-language tourist guidebook for Egypt was published in 1847, tourism and Egyptology have gone hand in hand. Travel historians Paul Starkey and Janet Starkey make a distinction between travellers – such as merchants, explorers, and antiquarians – and tourists – people who travelled for leisure.[40] Margarita Díaz-Andreu draws a similar line in her discussion of tourism for leisure versus tourism for educational purposes that would economically benefit the area. It is into the second category that Díaz-Andreu places what she calls archaeological tourism.[41] Archaeological tourism developed relatively late in the history of travel, around the mid-nineteenth century, and brought mostly the wealthy, leisured classes to places like Rome, Athens, and Cairo to see the ancient monuments for fun and to learn. It also had the benefit of improving infrastructure in those places that needed an injection of interest and outside money so the monuments could be maintained or restored. Early travellers to Egypt, such as George Sandys, Dominique Vivant Denon, Robert Hay, and John Gardiner Wilkinson, were not tourists in the strictest sense, even though they went to Egypt to view, see, gaze, and experience the country, its culture, and its ancient and modern monuments.[42] Their journeys were partly for leisure, but they also brought home records of their trips in the form of maps, diaries with social and political commentaries, drawings of monuments and tombs that have since disappeared, and more. For the purposes of this book, I will use the terms 'traveller' to mean anyone who travels, including archaeologists, and 'tourist' to mean someone travelling for leisure. Many, but not all, of the travellers in this book arrived in Egypt as tourists. That is, they first went to Egypt for purposes of leisure. Many of these tourists became travellers later on because they either left Egypt as a collector or returned to Egypt to financially support archaeology. Every one of them saw Egypt as a tourist at one point or another on their trip. To visit Egypt as a tourist and see what they were supposed to see, though, they would need a guidebook.

Humphreys points out that Murray's *Handbook for Travellers in Egypt*, prepared by Egyptologist and early traveller John Gardner Wilkinson, was based on the earlier *Modern Egypt and Thebes*, but with a new focus on the scholarly amateur traveller.[43] As Derek Gregory has argued, travellers and tourists passed through much of Egypt, and Alexandria specifically, as though it were an exhibition 'scripted by guidebooks'.[44] Baedeker's, Murray's, and Thomas Cook's guidebooks travelled 'ahead of and alongside the tourist' in order to establish the significance of sites and revealed 'an objective geography'.[45] Of the various guidebook options, including Murray's, Cook's, or Usborne's, I have chosen to rely primarily on Baedeker's *Egypt*.[46] Although earlier guidebooks existed, Baedeker's was one of the most widely used among tourists at this time.[47] It was so popular that some travel guides even marketed themselves as alternatives to Baedeker's, giving off-the-beaten-track advice and using titles such as *Travels Without Baedeker*.[48] But tourists were duly warned by other travel writers: 'To travel without Baedeker was not only to risk not recognising [an important site] when you saw one but to chance unfamiliar social and cultural encounters.'[49] Like many other books, Baedeker's guides imparted cultural information and practical advice on matters such as currency and the locations of physicians and the embassy, but it also contained more and better maps and a user-friendly starred rating system that encouraged many to depend on it. Moreover, I needed to rely on a guidebook that endured throughout the period covered by this study. Baedeker's *Egypt* ran to eight editions bound within sumptuous red leather covers over the five decades from 1878 to 1929, allowing me to trace changes in advice for travel activities in the area. For me, as for tourists and archaeologists at the time, it quickly became what one reviewer called 'the essential red coat'.[50]

With all the guides available in the nineteenth century, there was no shortage of travellers to or travelogues about Egypt. To understand what it was travellers were doing at the time, I rely on memoirs by writers such as Mark Twain, Amelia Edwards, Helen Mary Tirard, and others. Readers will travel through time and space in Egypt, much as an archaeologist or tourist would do. I present the experience of the sounds and sights of Egypt, much as might be done through a travelogue. The chapters partially present the experience of Egypt as an archaeologist – arriving in Alexandria, transferring to Cairo, seeing the pyramids and Cairo Museum as an expert might, staying in the hotels and attending dinners and meetings with other archaeologists and tourists, then making the long trip to Luxor. Here, they would think and talk, building their theories and working groups before heading out on site. There was always talking on site – around the evening fire to finish up the day and test out ideas. But it was in the urban hotels that archaeologists really started their work.

Organisation of this study

The idea for this book originally occurred to me when I was living in Cairo in 2010 and 2011 and teaching at the American University in Cairo (AUC). The country, and especially downtown Cairo's Tahrir Square, by the Nile, was in the midst of revolution and uncertainty. Yet there stood the Ramses Hilton Hotel, just steps from Tahrir Square, the centre of the revolution. Next to the hotel and Tahrir sat the recently burned out National Democratic Party headquarters, since torn down. Both the hotel and the party headquarters were tall buildings – one housing visitors and dignitaries visiting the city, the other housing a political system that Egypt's people hoped would end. Both were, without a doubt, political spaces.[51] Yet, by the middle of 2011, one remained standing, virtually untouched. The tall, brown tower with the bright blue Hilton logo at the top is a symbol of the continuing Western influence in Cairo, even after a century of independence. The juxtaposition of these two buildings, each representing a different political ideal, brought to my mind the old Shepheard's Hotel, which, until burned down in the Revolution of 1952, stood mere steps from the verdant, tourist-filled Azbakeya Gardens. From the 1870s until the 1950s, Western Egyptologists and tourists flocked to Shepheard's and the nearby Gardens. The strong association of both spaces to the British colonial presence in Egypt made the hotel one of the first casualties in the revolution that eventually pushed the British out for good. As a historian and Egyptologist, I wondered how people, especially Egyptologists, used that luxurious, yet political, central space. Most of the old Egyptian hotels are now gone, so I had no idea if enough evidence survived to answer my questions; I soon found that there was plenty.

That same year, Andrew Humphreys published his sumptuously illustrated *Grand Hotels of Egypt in the Golden Age of Travel.*[52] Humphreys, a travel writer himself, had done deep archival work in order to present a detailed portrait of these hotels in the late-nineteenth and early-twentieth centuries as tourist havens. He determined who visited Egypt, what significance the country held for them, where they wanted to stay, and why. Humphreys' work inspired mine. Excepting a brief mention of the fact that famous archaeologists like Howard Carter and his patron, Lord Carnarvon, stayed at Cairo's Continental-Savoy and Luxor's Winter Palace Hotels, Humphreys did not address Egyptology or Egyptologists in his book. His work showed me, however, that much archival material remained with which I might begin a study of these hotels and the archaeologists who stayed at them from around 1885 through 1925. Much like Humphreys, and the many others who write about travel in Egypt, my work follows the path taken by travellers up the Nile.

This book is organised geographically, as those who travelled through Egypt would have experienced it. The first chapter investigates Alexandria, the city that welcomed most Europeans and Americans upon their first arrival in Egypt. Alexandria was then, as it is now, largely ignored by visitors. Archaeologists and tourists alike usually stayed for just one night while waiting for the train to Cairo, and many left the city out of their stories except to say that they saw Pompey's Pillar upon landing and then spent the night in a usually unnamed hotel.[53] In the early days, the centre of European Alexandria was the Place des Consuls, which was the location of the two most popular European hotels, Rey's – favoured by the British – and Columb's – a favourite of the French.[54] After the British occupied Egypt in 1882, and tour guide Thomas Cook brought European tourists more regularly, hotels that catered to Europeans began popping up: the Savoy, the New Khedivial, The Hotel Abbat, San Stefano, and the Beau Rivage.[55] Such high-end hotels, where archaeologists on small budgets would not have stayed even for just one night, played a marginal role in the stories in this book. They were meant as long-term resort hotels where vacationing Europeans could spend the winter.[56] As E. M. Forster wrote, so many tourists and archaeologists themselves believed until the turn of the twentieth century that '[t]he "sights" of Alexandria are in themselves not interesting'.[57] But many of them toured Alexandria for want of something to do as they waited for their transport to Cairo, and to ease themselves into 'Oriental' Egypt. As more Greco-Roman Egyptian discoveries came to light, some tourists extended their stay in Alexandria for two or three days before moving south, to Cairo. The first chapter briefly traces the development of Greco-Roman archaeology in Alexandria and how those discoveries impacted travel in that city. Chapter 1 is more a history of archaeological tourism in Alexandria than of hotels as central sites. This discussion is included because Alexandria, a city that welcomed countless travellers to Egypt, was an important space that is so often ignored.[58]

Chapter 2 focuses on Cairo. Some visitors stayed the whole season in Cairo, venturing out for day trips to nearby Helwan and Memphis, but mostly staying around the city. Many archaeologists would spend several days or weeks in Cairo, preparing their equipment and making final preparations to go into the field for months at a time. When Flinders Petrie first arrived in Cairo, he stayed in the Hotel d'Orient, near the train station, because he could get room and board for 5 francs a day.[59] He then moved to the tombs at Giza, to save money. When Petrie had secured a university position as well as some more reliable funding for his excavations, he frequented the Hotel du Nil. Humphreys describes this hotel as one 'located at the heart of the native quarter, [where] a large and dark arched gateway gave into a courtyard filled with palms, and banana and orange trees, while the hotel building was a converted Arab house with striped stonework and *mashrabiya* screens'.[60] By the time Petrie stayed there in 1897, with his new wife Hilda at the start of their honeymoon excavation season, a rooftop tower had been built so guests could have views of the city.[61] It had a fairly central location, not too far from the Azbakeya, then an open, grassy area surrounded by other European hotels.[62] There were many European hotels near the Azbakeya, the best-known being the Continental, Shepheard's, the Savoy, and the Eden Palace. Most archaeologists at the end of the nineteenth century could not afford to stay in these hotels, which were designed for long-term tourists and cost more than their small excavation budgets would allow. A few, however, had very generous patrons. Early on, wealthy excavators like Charles Wilbour, Emma Andrews, and Theodore Davis could afford to stay at the world-renowned Shepheard's Hotel, where they could conduct business, such as hiring their crew, gathering supplies, and preparing their work plans, while their travelling companions shopped and saw the sights. Others, such as James Breasted and Howard Carter, had wealthy patrons who were willing to put their archaeologists up in the Continental, close enough to Shepheard's to walk over to sit on the famous terrace for tea and a talk. Other popular lodgings were pensions, private apartments or houses, and sometimes government residences. Chapter 2 introduces a lot of the characters in this book, and as Cairo was the city in which archaeologists prepared themselves, built their scientific networks, and readied their thoughts for the next step in their work, the chapter performs the same role for the argument in this book. That is, it works to introduce many examples and demonstrates the use of hotels as important nodes of networking, building the cognitive landscape, and being useful for certain knowledge-creating activities.

Having finished preparing for the season, some archaeologists went out to the desert areas near Cairo, and throughout Lower Egypt. But many went south to Luxor, heading up the river by steamboat, dahabeah, or train, and sometimes stopping at various points along the way. The third chapter follows these river travellers and centralises their activities on these semi-private boats as scientific institutions in Egyptology. The boats served as labs, classrooms, offices, storerooms, and homes. Some archaeologists, like Wilbour, Andrews, Davis, Archibald Sayce, and James Breasted, travelled to Luxor in dahabeahs, or private houseboats. They would live on the river in these floating homes, entertain guests, host scientific meetings, and even store artefacts to keep them safe. While dahabeahs were not necessarily options for all archaeologists on limited budgets, there were enough of them to analyse the role they played as semi-public spaces and as scientific institutions. James Breasted used dahabeahs in this manner, deliberately beginning to do so in 1905 and then continually after that for the next thirty years. He saw these floating laboratories as so important to Egyptology that he attempted, but failed, to get funding for a custom-built steamer to house his work in Egypt. Travelling up the Nile in any conveyance strengthened the bonds in each network, and, by turning the dahabeahs and steamers into scientific institutions themselves, they became truth spots by giving credibility to the work the travellers were doing.

The fourth chapter uses the sites and spaces in and around Luxor as the culmination of not only the long journeys of the travellers in this book, but also of the ideas presented throughout these pages. Truth spots, sites of knowledge creation, network creation, the intellectual landscape all peaked in the activities of archaeologists in Luxor. Luxor's many sites, tombs, and artefacts drew both archaeologists and tourists and, therefore, offered a variety of lodging options. Smaller hotels like the Grand Hotel, Karnak

Hotel, and the Savoy, were significant only as meeting places for social events and holiday meals, such as those that took place on Christmas Day, Boxing Day, or New Year's Eve. From the time it was built in 1907, the Winter Palace became the chosen lodging of many archaeologists, including Carter and Breasted before their dig houses were built, and Davis when he wanted to get off his boat. The Luxor Hotel, older than the Winter Palace and only a short walk away, was favoured by less generously funded archaeologists and tourists on a budget. This chapter is much longer than the others because there are two major hotel sites to discuss, and, because there exist far more sources for these events, the stories are more complex. Margaret Benson and Janet Gourlay made the Luxor Hotel their scientific centre when they excavated the Temple of Mut at Karnak from 1895 to 1897. Howard Carter's work on the tomb of Tutankhamun from 1922 to 1925 centred in his rooms and James Breasted's rooms at the Winter Palace Hotel. The common thread between those two stories is the work of Theodore Davis and Emma Andrews, who centred their decade of work in the Valley of the Kings on their dahabeah, but also in the Luxor Hotel and the Winter Palace Hotel.

The story presented in these pages ends around 1925, when the dispute Carter had with the Department of Antiquities over excavations at the tomb of Tutankhamun was ending. By this time, the antiquities laws that had allowed almost unabated excavation and the expatriation of artefacts had become much stricter. Laws were set by the newly independent Egyptian Government that no longer benefitted Western, rich, white, male excavators but ensured instead that Egypt would retain control over their own artefacts. For years, Egypt had fought for political and economic independence; and by 1922, after the First World War had changed the world order, the British had given them some autonomy. It was in 1922 that Carter found King Tutankhamun's tomb and all the 'wonderful things' it held. The control he tried to maintain over the artefacts he uncovered depended on his use of the space at the Winter Palace and drove the change in laws regarding archaeological finds. Luxor was the place in which, for this book, most of the work was performed and, therefore, was the most exclusive in terms of location and participation.

It is in these liminal yet enduring spaces that social and professional connections between and among Egyptologists will become clear. The spaces endure because the relationships that began here lasted long after people left hotels and Egypt for the last time. The work Egyptologists did in the social spaces of hotels reveals more about how scientific inquiry is done in the field: it has never been all about the trowel, the artefacts, or the subsequent reports. Egyptology is social; participation is decided in the spaces in which power is exercised. Building social and scientific networks and creating knowledge about the ancient world in hotels in Egypt demonstrates that the social *is* political, and that the remaining intellectual scaffolding leaves a problematic legacy.

Notes

1 W. M. Flinders Petrie, 'A digger's life', *The English Illustrated Magazine* (March 1886): 440–1.

2 *Ibid.* In fairness to Petrie, staying in Cairo and working at Giza in the early 1880s would have been very inconvenient until roads were paved from the city to the Pyramids and public tram service began. These modern additions cut the travel time, from a day on foot or several hours on a donkey, down to about 45 minutes.

3 Dena Goodman, *The republic of letters: A cultural history of the French Enlightenment* (Ithaca: Cornell University Press, 1994); Anne Secord, 'Science in the pub: Artisan botanists in early nineteenth-century Lancashire', *History of Science* 32 (1994): 269–315; Steve Pincus, '"Coffee politicians does create": Coffeehouses and Restoration political culture', *The Journal of Modern History* 67:4 (1995): 807–34.

4 Ann B. Shteir, 'Botany in the breakfast room: Women and early nineteenth-century British plant study', in *Uneasy careers and intimate lives: Women in science, 1789–1979*, eds Pnina Abir-Am and Dorinda Outram (Rutgers: Rutgers University Press, 1989), 31–44.

5 Kathleen Sheppard, 'Margaret Alice Murray and archaeological training in the classroom: Preparing "Petrie's Pups"', in *Histories of Egyptology: Interdisciplinary measures*, ed. William Carruthers (London: Routledge, 2014), 113–28; Alice Stevenson, *Scattered finds: Archaeology, Egyptology and museums* (London: UCL Press, 2019).

6 After Martin J. S. Rudwick, *The great Devonian controversy: The shaping of scientific knowledge among gentlemanly specialists* (Chicago: University of Chicago Press, 1985). On cognitive topographies in geology, astronomy, and other quantitative sciences, see, for example, David Phillip Miller, 'Method and the "micropolitics" of science: The early years of the Geological and Astronomical Societies of London', in *The politics and rhetoric of scientific method*, eds John A. Schuster and Richard Yeo (Boston: D. Reidel, 1986), 227–57.

7 Rudwick, *The great Devonian controversy*, 425.

8 *Ibid.*

9 Stephen Hilgartner, *Science on stage: Expert advice as public drama* (Stanford: Stanford University Press, 2000).

10 There are a number of ways to define Egyptology. Largely, it is the study of ancient Egyptian script and history, but I also include the practice of archaeology here. It is arguably impossible to study the script and history of an ancient culture without finding and attempting to understand the material culture. Throughout the book, Egyptology will include archaeology. For histories of archaeological

tourism, see, for example, Margarita Diaz-Andreu, *A history of archaeological tourism: Pursuing leisure and knowledge from the eighteenth century to World War II* (Cham: Springer, 2019); Donald Malcolm Reid, *Contesting antiquity in Egypt: Archaeologies, museums & the struggle for identities from World War I to Nasser* (Cairo: American University in Cairo Press, 2015), 137–66.

11 I have begun some of this; for example, Sheppard, 'Margaret Alice Murray and archaeological training in the classroom'; 'Tea with King Tut at the Winter Palace Hotel', *Journal of History and Cultures* 10 (2019): 67–88; '"Trying desperately to make myself an Egyptologist": James Breasted's early scientific network', in *Communities and knowledge production in Archaeology*, eds Julia Roberts, Kathleen Sheppard, Ulf Hansson, Jonathan Trigg (Manchester: Manchester University Press, 2020), 174–87.

12 For example, Derek Gregory, 'Scripting Egypt: Orientalism and the cultures of travel,' in *Writes of passage: Reading travel writing*, eds James S. Duncan and Derek Gregory (New York: Routledge, 1999), 114–50; Elliott Colla, *Conflicted antiquities: Egyptology, Egyptomania, Egyptian modernity* (Durham, NC: Duke University Press, 2007); Donald Malcolm Reid, *Whose pharaohs?: Archaeology, museums, and Egyptian national identity from Napoleon to World War I* (Los Angeles: University of California Press, 2002); Reid, *Contesting antiquity in Egypt*.

13 For example, Margaret Drower, ed., *Letters from the desert: The correspondence of Flinders and Hilda Petrie* (Oxford: Aris & Phillips, 2004); Paul Starkey and Nadia El Kholy, eds, *Egypt through the eyes of travellers* (Durham, UK: ASTENE, 2002).

14 For example, Jeffrey Abt, *American Egyptologist: The life of James Henry Breasted and the creation of his Oriental Institute* (Chicago: University of Chicago Press, 2011); John Adams, *The millionaire and the mummies: Theodore Davis's Gilded Age in the Valley of the Kings* (New York: St Martin's Press, 2013); T. G. H. James, *Howard Carter: The path to Tutankhamun* (London: I. B. Tauris, 2001); John A. Wilson, *Signs & wonders upon Pharaoh: A history of American Egyptology* (Chicago: University of Chicago Press, 1964).

15 For example, Rosalind Janssen, *The first hundred years: Egyptology at University College London, 1892–1992* (London: UCL Press, 1992).

16 But there are some. For example, William Carruthers, ed. *Histories of Egyptology: Interdisciplinary measures* (London: Routledge, 2014); William Carruthers and Stéphane Van Damme, eds, *History of Science, Special Issue: Disassembling archaeology, reassembling the modern world* 55.3 (2017); Christina Riggs, *Unwrapping ancient Egypt* (London: Bloomsbury, 2014); Julia Roberts, Kathleen Sheppard, Ulf Hansson, Jonathan Trigg, eds *Communities and knowledge production in archaeology* (Manchester: Manchester University Press, 2020). Two articles that discuss knowledge production in dig houses specifically are Colleen Morgan and Daniel Eddisford, 'Dig houses, dwelling, and knowledge production in archaeology', *Journal of Contemporary Archaeology* 2:1 (2015): 169–93; William Carruthers, 'Credibility, civility, and the archaeological dig house in mid-1950's Egypt', *Journal of Social Archaeology* 19:2 (2019): 255–76.

17 Steven Shapin, 'Lowering the tone in the history of science: A noble calling', in *Never pure: Historical studies of science as if it was produced by people with bodies, situated in time, space, culture, and society, and struggling for credibility and authority* (Baltimore: Johns Hopkins University Press, 2010), 14.
18 Aileen Fyfe and Bernard Lightman, eds, *Science in the marketplace: Nineteenth-century sites and experiences* (Chicago: University of Chicago Press, 2007); David Livingstone and Charles Withers, *Geographies of nineteenth century science* (Chicago: University of Chicago Press, 2011).
19 David Livingstone, *Putting science in its place: Geographies of scientific knowledge* (Chicago: University of Chicago Press, 2003), 13. See also, Simon Naylor, 'The field, the museum and the lecture hall: the spaces of natural history in Victorian Cornwall', *Transactions of the Institute of British Geographers* 27:4 (2002): 494–513; Simon Naylor, 'Introduction: historical geographies of science – places, contexts, cartographies', *British Journal for the History of Science* 38:1 (2005): 1–12.
20 Livingstone, *Putting Science in its Place*, 41.
21 *Ibid.*, 20. David Livingstone, 'Science, site and speech: Scientific knowledge and the spaces of rhetoric', *History of the Human Sciences* 20:2 (2007): 71–98; Mary Terrall, *Catching nature in the act: Réamur and the practice of natural history in the eighteenth century* (Chicago: University of Chicago Press, 2014).
22 Livingstone, *Putting Science in its Place*, 13.
23 See Alon Tam, 'Cairo's coffeehouses in the late nineteenth and early twentieth centuries: An urban and socio-political history', PhD dissertation, University of Pennsylvania, 2018.
24 Molly Berger, *Hotel dreams: Luxury, technology, and urban ambition in America, 1829–1929* (Baltimore: Johns Hopkins University Press, 2011), 12–19.
25 *Ibid.*, 6.
26 Reid, *Whose pharaohs?*, 73. See Table 0.1 for conversion rates. The hostels that cost 8 piastres/day would be equivalent to £6.80/$9.20 today; Shepheard's would be equivalent to £42.50/$57.50 today.
27 Andrew Humphreys, *Grand hotels of Egypt in the golden age of travel* (Cairo: AUC Press, 2010).
28 *Ibid.*, 15.
29 Hilgartner, *Science on stage.*
30 Thomas Gieryn, *Truth spots: How places make people believe* (Chicago: University of Chicago Press, 2018).
31 *Ibid.*, 3.
32 *Ibid.*
33 Food historians argue that commensality, the act of eating together, especially in public, defines identity. See, for example, Paul Freedman, *Ten restaurants that changed America* (New York: Liveright, 2016).
34 Michael P. Farrell, *Collaborative circles: Friendship dynamics and creative work* (Chicago: University of Chicago Press, 2001), 11.
35 Pamela Jane Smith, *A 'Splendid Idiosyncrasy': Prehistory at Cambridge, 1915–50* BAR British Series 285 (Oxford: Archaeopress, 2009), 33–6.

36 M. E. Newman, 'The structure of scientific collaboration networks', *Proceedings of the National Academy of Sciences of the United States of America* 98:2 (2001): 404–9; W. Glänzel and A. Schubert, 'Analysing scientific networks through co-authorship', in *Handbook of quantitative science and technology research: The use of publication and patent statistics in studies of S&T systems*, eds H. F. Moed, W. Glänzel, and U. Schmoch (New York: Kluwer Academic Publishers, 2005), 257–66.
37 For example, James Secord, *Victorian sensation: The extraordinary publication, reception, and secret authorship of 'The vestiges of the natural history of creation'* (Chicago: University of Chicago Press 2000); Ruth Finnegan, ed. *Participating in the knowledge society: Researchers beyond the university walls* (London: Palgrave MacMillan, 2005); Fyfe and Lightman, *Science in the marketplace*; Janet Browne, 'Corresponding naturalists', in *The age of scientific naturalism: Tyndall and his contemporaries*, eds Bernard Lightman and Michael S. Reidy (London: Pickering and Chatto, 2014), 157–69.
38 *Ibid.*, 169.
39 There are many examples cited in this introduction and that will be used throughout the book. Edward Said, *Culture and Imperialism* (New York: Alfred Knopf, 1993); Timothy Mitchell, *Colonizing Egypt* (Cambridge: Cambridge University Press, 1988); Timothy Mitchell, *Rule of experts: Egypt, techno-politics, modernity* (Berkeley: University of California Press, 2002); Starkey and El Kholy, eds, *Egypt through the eyes of travellers*; Hartmut Berghoff, Barbara Korte, Ralf Schneider, and Christopher Harvie, eds, *The making of modern tourism: The cultural history of the British experience, 1600–2000* (New York: Palgrave MacMillan, 2002).
40 Paul and Janet Starkey, 'Introduction', in *Egypt through the eyes of travellers*, eds Paul Starkey and Nadia El Kholy (Durham, UK: ASTENE, 2002), vii.
41 Díaz-Andreu, *A history of archaeological tourism*, 13–56.
42 Chris Naunton, *Egyptologists' notebooks: The golden age of Nile exploration in words, pictures, plans and letters* (London: Thames & Hudson, 2020).
43 Andrew Humphreys, *On the Nile in the golden age of travel* (Cairo: AUC Press, 2015), 29.
44 Gregory, 'Scripting Egypt', 134.
45 *Ibid.*; Inderpal Grewal, 'The guidebook and the museum,' in *Home and harem: Nation, empire and the cultures of travel* (Durham, NC: Duke University Press, 1996), 85–130.
46 There are a number of Baedeker's editions of *Egypt* I will be using. See the list of abbreviations at the front of the volume for information on the citation style for these books.
47 Humphreys, *Grand hotels*, 18; Alex Hinrichsen, *Baedeker's travel guides, 1832–1990*, 2nd edn, transl. Åke Nilson (Bevern: Verlag Ursula Hinrichsen, 1991). In 1942, Baedeker's headquarters in Leipzig were bombed by the Royal Air Force, burning up everything and leaving nothing but the front façade, so we have no information on sales numbers or anything else from before 1945. See also Herbert Warren Wind, 'Profiles: The House of Baedeker', *The New Yorker* 51:31 (22 September 1975): 42–93.

48 Ardern G. Hulme-Beaman, *Travels without Baedeker* (New York: John Lane, 1913).
49 Michael Haag, *Alexandria: City of memory* (New Haven: Yale University Press, 2004), 15.
50 Humphreys, *Grand hotels*, 20.
51 Kevin J. James, A. K. Sandoval-Strausz, Daniel Maudlin, Maurizio Peleggi, Cédric Humair, and Molly W. Berger, 'The hotel in history: evolving perspectives', *Journal of Tourism History* 9:1 (2017): 92–111.
52 Humphreys, *Grand hotels*.
53 Drower, *Letters from the desert*, 118.
54 Humphreys, *Grand hotels*, 24–5.
55 *Ibid.*, 25–37.
56 Petrie stayed at the Hotel des Messageries in Alexandria in 1881. According to Robert Playfair's *Handbook to the Mediterranean*, 2nd edn (London: John Murray, 1882), the hotel was on the Rue de la Bourse, near the sea (50). It is possible it was a hotel associated with the French merchant company Messageries Maritime.
57 E. M. Forster, *Alexandria: A history and a guide* (Alexandria: Whitehead Morris Limited, 1938), ii.
58 See Toby Wilkinson, *The Nile: Travelling downriver through Egypt's past and present* (New York: Random House, 2014) in which he traces a wonderfully detailed journey down the Nile from Aswan, abruptly ending the book in Cairo.
59 W. M. Flinders Petrie, *Seventy years in archaeology* (New York: Henry Holt, 1931), 21. See Table 0.1. Today that is equivalent to about £17/$23 per night.
60 Humphreys, *Grand hotels*, 50.
61 *Ibid.*
62 *Ibid.*, 59.

1

Alexandria: archaeological tourism in a city forgotten

> I will spare the Gentle Reader descriptions of the journey and of the picturesque dirt of Alexandria.[1]

So said Amelia Peabody, the intrepid (if fictional) Egyptologist in her 1885 journal. Whether coming from the United States, England, or the rest of Europe, when travellers arrived in Egypt in the late nineteenth and early twentieth centuries, they usually arrived first in Alexandria, and had to make their way to Cairo by train.[2] Thinking Alexandria was too European a city – that is, not exotic enough for their expectations of Egypt – travellers tried to leave as quickly as they could. In the 1880s, archaeologists and tourists alike stayed for a few hours, or as much as one night, waiting for their train to Cairo. The city, its history, its hotels, and its inhabitants were largely passed over by both tourists and archaeologists in favour of locations further south.[3] There were some archaeologists who stayed to try to excavate the rapidly disappearing mounds of ancient detritus – like Heinrich Schliemann, David Hogarth, and Fred Benson – but not many returned after one season's failure at having found anything pharaonic.[4]

Alexandria was, and still is, a transitional city for a lot of travellers, historians, and archaeologists. Founded around 332 BCE, it was a relatively new city in Egypt. Because it was so new, it could not really be included in the pharaonic histories of Egypt. It was a city with Greek roots and Roman influence, but because it was in North Africa and therefore not as rich in archaeological finds as the sites of Athens or Rome, Alexandria was not usually considered part of Classical archaeology, either.[5] Stuck between two different continents and cultures, it was a city with a singular history, and in the late nineteenth and early twentieth centuries archaeologists were still trying to figure out its significance. For travellers, Alexandria allowed for a relatively easy transition from their European comfort zone to an Egyptian one.[6] The city had enough European influence in language, food, and clothing that it was a soft landing spot for novice travellers on the way to the more exotic city of Cairo. Yet, the climate, architecture, shopping, and people were Egyptian enough to ease visitors into the Eastern world without being overwhelming. For most travellers, it was a useful entry into Egypt, but they hurried past it to Cairo in search of the picturesque.[7] Using Baedeker's, Murray's, and Cook's guides, tourists and archaeologists alike were rushed through a curated Alexandria so they could get to the places and sites that the guidebooks spent much more time talking about. In fact, Baedeker's *Egypt* devoted almost nine times as many pages to Cairo than it did to Alexandria in the 1885 edition: Alexandria occupied 20 pages to Cairo's 177.[8] Over time this ratio remained steady even if absolute numbers varied; the book was meant to lead the traveller to Cairo and points beyond.

These travel patterns and the attitudes that bolstered them were keenly felt by Alexandrian residents. In 1922, archaeologist and director of the Greco-Roman Museum, Evaristo Breccia, was arguing that

> Nothing is more false than the widely spread idea that Alexandria has nothing to show to its visitors. This fiction has arisen from the fact that Alexandria is owing to its position, a point of arrival and a point of departure. The tourist arrives in Egypt eager to see the Pyramids and the grand ruins of Pharaonic civilisation whose description have stirred his imagination since childhood ... Alexandria, for him, is nothing but a port. But if he does not tarry, he will have but an incomplete idea of the marvelous history of this country, dead a hundred times and a hundred times resuscitated, and he will leave with a regrettable gap in the series of his impressions and his knowledge ...[9]

By 1929, thanks to Breccia and other archaeologists who had uncovered and made accessible more archaeological sites, some tourists were staying as long as two or three days. But even then, most Egyptologists did not stay longer than a day and a half, needing to move on to Cairo and the rest of their work upriver, in the Delta, the desert, and throughout Egypt.

Alexandrian archaeologists seek the remnants of the Greco-Roman period in Egypt, a specific field of study with relatively few practitioners but archaeologically important for a number of reasons. For the first 100 years or so of concentrated archaeology in Alexandria, investigation, organisation, and administration of the sites was dominated by Italian practitioners. Despite rapid and destructive urbanisation in Alexandria since the 1880s, Giuseppe Botti, Evaristo Breccia, Achille Adriani, and others unearthed remains of the ancient city and large tomb complexes that cover several acres. These sites and their artefacts told the long, rich, and exciting story of Alexandria, and gave more tourists more to see.[10] Currently, many of the surviving remains of the Ptolemaic city (c. 332 BCE–32 CE) are underwater in the port of Alexandria. They have lain there for centuries and, in order to study and preserve them, some objects are being brought up by maritime archaeologists while others remain submerged.[11] The archaeologists that are the focus of this volume did most of their meeting, talking, and network building with their colleagues not in Alexandria, but in Cairo and Luxor, and places in between. However, the events and discoveries in Alexandria that caught the attention of tourists in the nineteenth century were the same that kept scientists in the city for a few extra hours. No matter the purpose of their visits, upon entering Alexandria everyone essentially became tourists, especially on their first trip into the country; therefore, it is the historical tourist experience in Alexandria, closely tied to the archaeology, that is the focus of this chapter.

Like the city itself, this chapter about Alexandria will have a slightly different feel from the other places in Egypt and the other chapters in this book. This history of archaeological tourism in Alexandria is not quite an investigation of travellers' use of hotels in this city. Instead, this chapter aims to understand how travellers interacted with the city as they simultaneously tried to move through it quickly on the way to more exotic destinations. A brief history of Alexandria's archaeological tourism demonstrates the power of excavations and how new knowledge about the history of the city, presented to people through guided museum displays and publicly accessible sites, expanded tourism in the city from the 1890s to 1929.[12] Since travellers viewed Alexandria as a transitional space, archaeological findings were crucial to enticing European tourists to stay there longer. This chapter investigates past journeys through the city, seeing vicariously through visitors' memories from their journals, letters, and published memoirs. These crucial sources, which pair with volumes of Baedeker's *Egypt*, trace what tourists might have seen in Alexandria, from just before the opening of the Greco-Roman Museum in 1892 until the 1930s. Alexandria's timeline is a little different than the rest of the book, too, because of the influx of work leading to discovery from 1892 to 1929. These new discoveries meant that Baedeker's altered the exhibit of Alexandria, that is, they re-curated the recommendations for how much time tourists should spend and which paths to take while visiting the city. However, much like the Victorian domestic advice books that piled on family bookshelves during this period, guidebooks were meant as idealised versions of a journey.[13] Simply because a guidebook suggested an activity or path did not mean that the reader dutifully followed; therefore, tourists' accounts and archaeologists' memoirs are crucial in this chapter, in order to gauge real travel experiences.[14] All of this recommended tourist viewing was dependent on the context of the archaeological discoveries that were coming out of the ground. These new artefacts and sites drew attention and tourism to Alexandria, more so than the beaches, casinos, and sunny summer weather. By taking part in some archaeological tourism in Alexandria in the context of scholarly archaeological developments, this journey through the city will bring to light the important Greco-Roman sites in Alexandria that too often go unnoticed in traditional histories of archaeology in Egypt, and by archaeological tourists today. Although Alexandria still does not spark the same amount of tourist interest as Cairo and sites further south, the Greco-Roman period findings are central to seeing archaeology as a mechanism for attracting tourist attention.

While the hotels or other places of residence are not the main point of this chapter, Alexandria is still an important starting point for discussion. Many travellers omit Alexandria from their journals, letters, or memoirs.[15] Some more recent books about travelling on the Nile have also left it out.[16] Despite these gaping holes in traveller's written memories, Alexandria was a place where our travellers arrived, got their bearings, and began their journey in a new country and culture. Focusing on Alexandria will add to our discussion of archaeologists' activities by providing a new angle to the story of travel in the nineteenth and early twentieth centuries. E. M. Forster himself used Baedeker's guides to introduce himself to his beloved city during the First World War.[17] So, this path through Alexandria follows in the carefully recommended footsteps of thousands of famous and unknown travellers seeking the sights and sites of the past.

Baedeker's archaeology: a historical tourist in Alexandria

Most guidebooks opened chapters about each city in Egypt with the history of the city in order to orient travellers. Alexandria is the city of Alexander the Great, who, according to legend, dreamed of the city and its location before its founding. The city faces northwest, looking out over the Mediterranean toward Greece. Known as the 'megalopolis' of the Hellenistic world, people have lived in the city continuously since its founding in 332 BCE (Figure 1.1). Alexandria's population has risen and fallen drastically throughout its two millennia, but by the time the overland route to India was popular with Europeans in the 1840s, more people were visiting, and more people were staying overnight. Before the Suez Canal was finished in 1869, many travellers bound for India would disembark at Alexandria, proceed to Cairo by boat through the canals to the Nile River, then take a carriage (which was superseded by the railway in 1858) across the desert to Suez to board a ship for Bombay.[18] Sometimes travellers would stop in Cairo for a few days before going to Suez; it all depended upon the boat schedule. Alexandria was also the first port of call for anyone coming from Europe or the United States to spend the winter in Egypt. Depending on their arrival time, the time it took to get through customs, and the scheduled departure of their transportation to Cairo, some of these travellers stayed overnight in Alexandria.

By 1874, the city had a population of 270,000 people. In 1882, the British bombarded Alexandria's harbour in a successful but brutal squashing

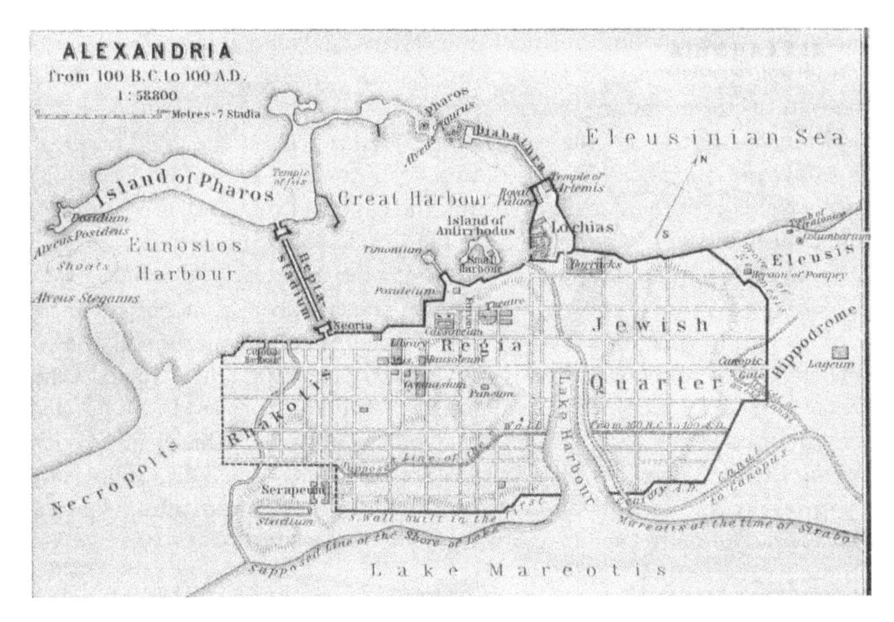

Figure 1.1 Baedeker's map of Alexandria, 100 BCE–100 CE.

of the 'Urabi nationalist uprising in order to reseat Tewfik, a leader who would cooperate with their presence.[19] Not only did the bombardment destroy major parts of the city and its Egyptian population, but it also alarmed a number of European residents to the point that they moved away. The travellers were quick to return. It is important to note that in the 1885 and later editions of Baedeker's *Egypt*, the uprising was discussed as 'the fury of the natives', and later as a 'misfortune', virtually erasing it as a defining political moment in Egypt's history that would affect the next 130 years and counting.[20] With the British having a 'paramount' influence in the country from the time they defeated 'Urabi and his forces, they established a safe place for Europeans to live under largely British control by 1897, when 46,000 of the 320,000 residents in Alexandria were European.[21] Americans were also visiting Egypt more as part of the European Grand Tour. In fact, by the end of the nineteenth century, it has been argued that travellers on the Grand Tour who did not go to Egypt would be looked at askance and their stories doubted.[22]

As early as 1844, William Makepeace Thackeray wrote that Alexandria was 'as exotic as the London port of Wapping', and that 'the curiosities of Alexandria are few, and easily seen'.[23] If that was the case, it is easy to see why many travellers tended not to discuss the city in their travel journals. Over three decades later, not much had changed. Even though she and her group arrived first in Alexandria, Amelia Edwards famously began her travelogue *A Thousand Miles Up the Nile* in 'the great dining-room at Shepheard's Hotel in Cairo during the beginning and height of the regular Egyptian season'.[24] She ended her journey there as well. She mentioned Alexandria a few times within the pages of her story if only to say that the first glimpse of the pyramids on the train from Alexandria, probably like her thoughts of the city itself, was 'not impressive'.[25] Correspondingly, in the 1878 and 1885 editions of their guidebook, Baedeker's *Egypt* described Alexandria as an 'Europeanised' city, and advised that the traveller could, 'by taking a carriage, inspect the town with its few relics of antiquity in half-a-day'.[26] Compared to the almost two-week itinerary given for Cairo during the same period, the guidebooks were pushing people south quickly.

There were a number of counter-points to this dismal view of Alexandria, however. Jenny Lane, travelling with Amelia Edwards and Lucy Renshaw in 1873, wrote of her first sighting of Alexandria on 27 November:

> I dont [sic] think I shall ever forget the beautiful torquise [sic] blue sea & the lov[e]ly old Town with its numberless Wind Mills in the distance; & Palm trees in the back ground. The Pilot a black Arab came by the side of the boat about 11.30. but would not come on board as we were all in Quarantine we arrived at Alexandria at 1. Oclock. and shall have to remain on the Boat till Saturday. A Yellow Flag being hoisted the Quarantine colour.[27]

Lane also enjoyed their time travelling through Alexandria for the day or so they were able to, but the women did not stay long.

Archaeologists themselves had pronounced their despair for studying Alexandria or for the city producing anything of interest to the discipline. Sometime in the 1880s, Heinrich Schliemann – infamous for excavating the site of Troy in the 1870s – came to Alexandria in hopes of finding Alexander the Great's tomb. True to his nature, Schliemann quickly dug a few trenches near the Ramleh tram station, where the Caesareum is supposed to have stood, and found some later Roman material remains. Disappointed, but probably not surprised, he left Alexandria behind and headed out for a leisure cruise on the Nile.[28] In 1895, Oxford archaeologist David Hogarth, along with Fred Benson, was sent by the Egypt Exploration Fund and British School in Athens to evaluate the potential at Alexandria as a site for Classical archaeology.[29] Hogarth decided that '[n]o foreign society, which can find almost virgin sites, could be invited to search in Alexandria for obscure Graeco-Roman ruins or bare topographical indications' because in Alexandria, the remains were under meters of rubble and the water table.[30] Scholars did not hold out much hope that much would be found under the rapidly expanding city.

As early as 1882, however, Murray's *Handbook* and Baedeker's *Egypt* had started to include instructions for travellers landing in Alexandria not just in terms of how to get through the arrival unscathed, but also how to stay and enjoy the city.[31] After travellers had prepared themselves on the journey over by reading about the history of the city, Baedeker's prepared them for the terrain. Upon arriving at the 'perfectly flat N.E. coast of Egypt' travellers would not immediately see the city until just before they entered the harbour.[32] Their eyes were usually drawn to the main features: the lighthouse, the Khedive's palace at Ramleh, and the Beduin Gate.[33] As the boat moved closer to the harbour, they would see windmills, and before 1880 they would see both of Cleopatra's Needles – one standing and one on the ground.[34] Then, as now, Pompey's Pillar was easy to find and was a 'convenient landmark' for travellers (Figure 1.2).[35] Hilda Petrie, wife of Flinders Petrie, briefly mentioned the city when she wrote home about arriving in Alexandria for her honeymoon excavation season in 1897. She wrote 'at last we are sighting the African coast, and Pompey's Pillar is the first thing we can see'.[36] She went on to talk about the chaos of landing, and getting through customs, writing to her family,

> oh! the landing! We took up our stand, with our bags, close to the square hole one emerges from, and then waited, and the whole Arabian Nights poured in upon us, pell-mell, in every wild richness of oriental dress – porters tumbling in for baggage.[37]

Following the advice for disembarking, getting through customs and to one's hotel was crucial to the start of a successful trip for first-time visitors.

Figure 1.2 Stereoscope of Pompey's Pillar, Alexandria, Egypt, Keystone View Company, c. 1899. Not actually belonging to anyone named Pompey, this column was erected in honor of Diocletian around the turn of the third century, CE.

After introducing the city itself, this was the next section in almost every guidebook.

Most European travellers to Egypt had never been out of Europe, so their first entry into Egypt was an experience unlike any they knew. After a 'brief sanitary inspection', travellers were to get off their boat and find a hotel porter, first and foremost, to arrange for the care and transport of their luggage, which usually came in the form of large trunks and boxes. Many times, travellers had the help of travel companies under which their trips were booked and organised; this included Thomas Cook and Son's, Clark's, Gaze's, or the Hamburg-American lines.[38] The guidebooks were clear on how to handle Egyptians, who were considered to be not as civilised as white Europeans. When Margaret Benson and her brother Fred were travelling with Thomas Cook in 1894, a year before Fred excavated with Hogarth, they followed Cook's instructions when they arrived in Alexandria. Benson wrote to her mother that, when they landed, all the Egyptians,

> rushed and offered one boats and hotels and carriages *ad infinitum*, and to all Fred only said, 'I want Cook.' Then at last a man dressed apparently in a white nightgown and scarlet jersey, with 'Cook' in flaming yellow embroidered on him, hurled himself into Fred's arms – then we felt safe, and whenever we got offered anything else, we only said, 'I've got Cook.'[39]

But there was hope, with Baedeker's informing their readers that 'if the traveller makes due allowance for their shortcomings, and treats the natives with consistent firmness, he will find that they are by no means destitute of fidelity, honesty, and kindliness'.[40] Not much had changed between the 1908 and 1914 guidebooks, which warned visitors that the 'mere children' had a 'touching simplicity' but could also be shrewd exploiters for their own financial gain.[41] This was the way that many Britons at the time characterised indigenous groups throughout the parts of the world they controlled: too childlike and ignorant to govern themselves, but also intelligent, cunning, and shrewd when it came to business and finance. These colonial attitudes impacted relationships between travellers and Egyptians in myriad ways, as this and further chapters will show.

Seeing the sights

If travellers were able to catch their train right away and were not staying in Alexandria overnight, the road to the station was all of the city the traveller would see. Guidebooks knew and could also dictate travellers' patterns, so they supplied some basic information about sites along that particular route. But if visitors had a half day or more to spend, Baedeker's and other guidebooks proposed and altered a number of itineraries through

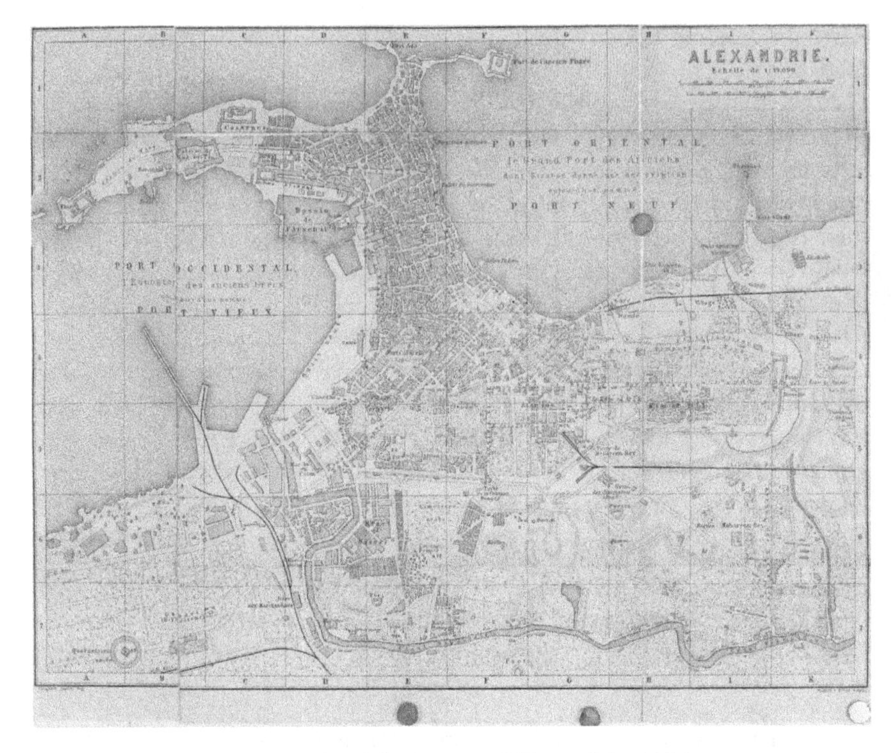

Figure 1.3 Baedeker's map of Alexandria, 1885.

the city over the years (Figure 1.3). As a European-esque town, there were several European shops and theatres for passing the time in a more comfortable manner for the traveller; however, guidebooks did not really focus on these activities, and instead implored the traveller to see Alexandria. From the mid-nineteenth century, European Alexandria was centred in the Place des Consuls, where the two most popular European hotels, Rey's and Columb's, were located.[42] Some Americans stayed in the Hotel Abbat, which was 'well situated' in town on St Catherine's Square.[43] It is difficult to say where Schliemann, Hogarth, and Benson stayed when they were excavating in Alexandria. It is clear from their letters home that Margaret and Fred Benson stayed at the Hotel Abbat when they arrived. Further, toward the end of the nineteenth century – once Thomas Cook brought more European tourists, after the British fully took control of Egypt — the Savoy, the New Khedivial, San Stefano, and the Beau Rivage were built and became popular because they catered to European clientele.[44] These hotels could not compare to the luxury of some of the hotels in Cairo, but they were deemed 'very fair' hotels, at least.[45] It is unlikely that these high-end hotels were places where archaeologists on small budgets would have stayed just one night, with the possible exception of the New Khedivial, which was located right by the train station. For the most part, the hotels in Alexandria were meant as long-term resort hotels where vacationing Europeans could spend the winter, and wealthier Egyptians might escape the heat of Cairo in the summer.

The infamously frugal Flinders Petrie stayed at the Hotel des Messageries in 1881, and it is not hard to see why: it was affordable and in the area of town frequented by locals.[46] In the 1878 Baedeker's *Egypt*, the Hotel des Messageries was regarded as a second-class hotel, located 'in the Okella Sursok ('okella' being the name for a single large block of houses) ... on the New Harbour, on the E. side of the town; charge 12 ½ fr. including wine; recommended to persons of moderate requirements'.[47] Robert Playfair's *Handbook to the Mediterranean* described the hotel as being on the Rue de la Bourse, near the sea.[48] It may have been a hotel associated with the French merchant company Messageries Maritime. This company not only shipped goods and mail on the Mediterranean, but they also had berths for people travelling to Egypt. It is possible that the hotel was meant for guests on the Messageries ships for their overnight stay in the city before leaving for Cairo or for their boat back home. Petrie was not known for keeping a detailed journal of his personal activities until his wife Hilda joined him in 1897, but he did write journal-like letters back home. It is, however, unclear exactly where he stayed when he was in Alexandria, but he said he was well-fed (by his own standards, which were problematic to begin with) and could sleep. By the 1885 edition of Baedeker's *Egypt*, the Hotel des Messageries had disappeared from the hotel options, either because it had fallen into such disrepair that it was no longer used by Europeans or because it had been destroyed by the British bombardment of Alexandria three years earlier. In 1897, the Petries had no intention of staying in the city, so they had no hotel plans. According to Hilda's letter to her family, the couple disembarked with their luggage and 'drove straight to the station, but it was already too late for the aftn. train, so we got back to a hotel ...' but she did not record which one.[49] Since they were stuck in the city for a night, Hilda wrote that the pair 'pottered about the native quarter', but then hurriedly moved on to Cairo the next day, without stopping to see any sights.[50]

American Egyptologist James Breasted was known for keeping meticulous records in the form of letters home to his family.[51] On his first trip to Egypt in 1894, his honeymoon with his new wife Frances, James remarked in a letter home to his parents that he had spent the better part of the Mediterranean journey from Italy 'devoted to a vigorous study of the guidebooks on Alexandria'.[52] Although he did not say which books he used, he may have had an 1891 edition of Murray's *Handbook* on this trip. The couple was coming from Berlin, so Breasted was probably using a Baedeker's *Egypt*, which would have been the 1885 edition; the Cook's edition would have been from 1875 or 1888. Later on, Breasted owned an 1898 edition of Baedeker's. The 1891 Murray's *Handbook* and the 1898 edition of Baedeker's were in his personal library, which was left to the research archives collection of the Oriental Institute.[53] He probably followed the detailed guidelines for coming into the city and getting to a hotel, and then decided to follow the well-worn path to Pompey's Pillar before heading back to the train station.

The couple landed in Alexandria the afternoon of 30 October and spent much of the rest of the day getting themselves and their cases off the boat and to the hotel. They had the aid of Gaze's dragoman, or guide, who 'brought us safely through the struggling, sweating, hauling mob'.[54] Too late for the afternoon train, the couple arrived at their 'little French Hotel' by 9:00 pm, which may have been Columb's.[55] They wrote that they 'were very well served indeed', slept under a mosquito net for the first time, and planned to take the morning train to Cairo at 9:00 am.[56] It is likely, while resting in their hotel, that the Breasteds heard the sounds of the city which, according to early tourists, consisted of donkeys braying, camels grunting, and stray dogs barking long into the night.[57] Murray's suggested their readers should take a stroll through the bazaars, just as the Petries would do in 1897, which they touted as 'almost the only Oriental feature of the city'.[58] The Breasteds, however, did not have time to wander. Before the couple went to the train station the following morning, they allowed for about an hour so they could 'have a look at Pompey's Pillar, which,' Breasted wrote, 'is the only monument in Alexandria'.[59] On their arrival, they saw that the sand on the hill was full of residential huts, and that the hill was 'not so bare as it seems at a distance ...'.[60] Due to the living conditions near the pillar, Breasted told his family back home that, as he was readying his camera to take a photograph of Frances near the base of the pillar, 'a young woman came & squatted down behind me & I quite understood why the hill is not so "bare"'.[61]

None of the guidebooks that Breasted possessed on this trip would have mentioned the Greco-Roman museum, founded in 1892, so other than describing the Pillar and the short amount of time they spent there taking a photograph, he moved on to describe the 'long, hot dusty ride up the Delta from Alexandria to Cairo'.[62] As early as 1894, some travellers added a visit to the Catacombs at Kom es-Shukafa, and, until 1880, to Cleopatra's Needles; others would make it all the way out to 'the rubbish heaps of the ancient *Nicopolis*', well east of the city.[63] But for the most part, the Breasteds took the traditional route into and out of Alexandria that was popular with visitors for almost fifty years.

Then everything changed. After the third edition of Baedeker's came out in 1895, and the fourth edition in 1898 which joined Upper and Lower Egypt into one volume, tourists were advised that they could see the sights of Alexandria in one or two days, instead of just a half a day. There were more European hotels listed, though they were, as the guide was quick to point out, 'below the standard of the Cairo hotels'.[64] More available hotel rooms meant that more people were coming to Egypt in general and staying at least one night in Alexandria before moving on. Hotels Khedivial and Abbat were still listed along with an increased number of second-class hotels like Hotel du Canal de Suez and Hotel des Voyageurs.[65] Tourists were now encouraged to devote more time during their stay to the historical sites of the town and not just pass them by, even though many continued to ignore them on their way from the boat to the train.

By 1898, the first day in town was for seeing the 'semi-Oriental' city itself.[66] As a part-European/part-Egyptian city, travellers were duly warned that it would be 'unadvisable to postpone the inspection of Alexandria until the return, for by that time the traveller is saturated with other impressions'.[67] Sightseers were advised to drive through the town and bazaar and to allow time to visit the Ras el-Tin Palace. In the afternoon they were to go to the then-suburb of Ramleh to see the villas and return by the Mahmudiyeh Canal. The second day in town was to be devoted to antiquities: the museum in the morning and Pompey's Pillar and associated excavations of Dr Botti in the afternoon. By this point, it was clear that the presence of the Greco-Roman Museum, and the central place it took in Baedeker's guidebooks, played a role in shifting the advice for how tourists should see and experience Alexandria.

In the late nineteenth century, arrangements to see these sites were to be made with Botti's office at the museum. In 1895, the Greco-Roman collections were described as being displayed in 'The house known as "Kirkor", in the E. part of the Rue de Rosette', and they were in the process of building a purpose-built museum on the same street.[68] Just three years later, in 1898, the new Greco-Roman Museum building was included in the short Alexandrian itinerary. Collections in the museum came largely from Botti's excavations, in addition to many Greco-Roman Alexandrian pieces that had been returned from the Egyptian Museum at Giza in order to complete the collection. Botti's excavations and the museum he built around them firmly established Alexandria as the centre of Egypto-Greco-Roman antiquities. The artefacts, like the city itself, were fully Greek and Roman as well as fully Egyptian; there was nowhere else in the world like it. At this unique collection, tourists could view the history of eras that other museums did not include or that seemed artificially tacked on at the beginning or end of other periods. Baedeker's *Egypt* presented a full guide to the museum in the 1895 edition, walking the tourist through the museum, room by room, including a detailed item by item description.[69] In Baedeker's 1902 edition, the Greco-Roman Museum still had the detailed guide, but the name of the street the museum was on had changed from Rue de Rosette to Rue du Musée.[70] Rue de Rosette still existed, but the part of the street the museum was built on, which 'diverges to the N. from the Rue de la Porte Rosette', received a name change, reflecting the importance of the museum to the city, to tourists, and to scholarship of the area.[71]

Even though the museum guide occupies only about two pages in total (compared to the Cairo Museum's twenty-five pages), there is so much detail even present-day readers could use the guide to find their way through part of the collections. The description begins in the vestibule, which contained a 'Colossal marble statue of Hercules; two capitals with basket-work ornamentation from the old church of St. Mark at Alexandria'.[72] The guide then walked readers through the sixteen rooms within the museum, which contained items such as late-period mummies, tombstones of Roman soldiers, marble statues of emperors, and some statues of pharaonic figures like Ramesses II. The garden also contained 'a number of larger monuments, including limestone and porphyry sarcophagi, fragments of Greek columns, a red granite group of Ramses II and his daughter (from Abukir), etc'.[73] By 1914, the museum had grown to twenty-three rooms and a garden, in order to accommodate the artefacts from the excavations of Kom es-Shukafa and the rock-cut tombs. The public reports on these excavations appeared in the form of oral presentations Botti gave to the Société d'Archéologique d'Alexandrie, making them hard to find in print. There are, however, some brief reports in the *Bulletin de la Société d'Archéologique d'Alexandrie*.[74] Botti was the main, if not the only, author of many of these bulletins while he was the director of the museum. Many of the reports are about artefact assemblages, but not about specific sites. From 1898 to 1902, Botti included a short section called 'Additions au 'Plan de la Ville d'Alexandrie, etc.' in which he outlined the 'Fouilles de la Direction générale des Antiquités', or the Excavations of the General Office of Antiquities.[75] The 1898 publication was the first report of this kind, so his excavation notes summarised developments all the way back to 1892.[76] He discussed the museum itself, the administration of the museum, and the impact of the incoming artefacts. The volumes exist up to the year 1912, when possible budget cuts and, later, the First World War stopped them. What travellers needed to know was that the museum was growing quickly, adding dozens of major artefacts to the collections every year. The growing collections increased the public's interest in touring the museum and the sites themselves. Tourists were learning that Alexandria was an important historical site that deserved some of their attention.

Once visitors had spent their morning at the museum, they usually left with permissions to visit sites in hand. Beginning in the Place Mehemet Ali, visitors would ride or drive to Pompey's Pillar (column) first, then to see the catacombs near it, then to the tombs nearby, discovered by Botti in 1893. The 1895 edition told visitors that the tombs were 'highly interesting' but did not convey to visitors the importance of these early sites mainly because they were not yet fully known or understood. By the 1898 edition, however, Baedeker's included more, and more thorough, description of sites, likely based on Botti's brief excavation reports. By 1908 it cost three piastres for travellers to enter the complex for Pompey's Pillar.[77] Travellers were directed to climb the flight of stairs to the top of the plateau on which the Pillar stood. From this vantage point, one could look around and see 'fragments of Roman buildings and other objects brought to light by the extensive excavations begun by Botti ... and lately recommenced by Dr. Breccia'.[78] Reading about the pillar, travellers learned that, made of red granite from Aswan, the height of the pillar itself was sixty-eight feet; with pedestal and column topper it was eighty-eight feet high. The west side of the column had an inscription from 392 CE; to the north of the pillar was an ancient water basin.[79]

Moving further on, and also costing three piastres, the underground passageways of the Serapeum were apparently 'of little interest' to the regular traveller in 1908.[80] By 1929, however, there were passageways and chambers with 'small niches of unknown purport' inside.[81] The niches, which travellers can still visit, were carved from grey granite, with no decoration at all. After Botti's excavations at the Serapeum and Pompey's Pillar from 1892 to 1898 had produced a massive black basalt Apis bull that still stands in the Greco-Roman Museum, a German expedition excavated at Pompey's Pillar from 1898 to 1902, but without major artefacts to draw the tourist's attention. After Botti's death in 1904, Evaristo Breccia reconvened the excavations there and added stairs for ease of public access. As the Pillar is still one of the most visible monuments in the city, it has continued to draw visitors to the entire complex surrounding it.

One major change in the 1902 edition of Baedeker's was a discussion of the excavations that Botti and a local physician Dr Schiess Bey were carrying out in what was then called the Egyptian Burial Place, or the catacombs at Kom es-Shukafa. In the first edition of Baedeker in 1878, tourists were told about the catacombs, but it was taken for granted that the tombs would not exist much longer because of the destructive quarry work being done nearby.[82] Thanks to Botti's quick excavations, the 1902 edition presented a full discussion of the main large tomb. It was discovered in September of 1900 when, according to the story, a donkey fell through a hole in the ground and into one of the tombs.[83] The tomb held material remains that dated from the second century CE, and, as Baedeker's informed their readers, was 'an admirable example of the characteristic Alexandrian fusion of the Egyptian and Graeco-Roman styles'.[84] The work at the catacombs of Kom es-Shukafa was completed by the end of 1904, the year Botti died. In their June 1904 proceedings, the Society of Antiquaries of London reported on the excavations and how in just four short years, Kom es-Shukafa had been cleared completely, and was now lit with electric light. Supplying lighting was no small feat given the size of the site and the difficulty in clearing quarried areas, not to mention the modern city encroachment. The Society of Antiquaries also commended the late Botti on the thoroughness of his work.[85] Baedeker's included a map of the tombs and, true to form, a room-by-room, self-guided walking tour, which now highlighted the Sepulchral Chapel where the sarcophagi were cut from the solid rock.[86]

By the start of the First World War in 1914, the site was known as the Catacombs of Kom es-Shukafa, meaning the hill of potsherds. It became, and remains, the most significant known burial site in the city, with its many layers of chambers and exquisite decoration. The map of the site from Baedeker's was extremely detailed, showing the staircase that would take visitors down the two stories to the bottom of the complex, but the lower level was usually under water.[87] Very much like the museum guide, Baedeker's pages detailing the tombs were meant to walk visitors through the site as a tour guide, explaining not only the structure but also its meaning. Many descriptions would still be informative for today's visitor:

> The façade of the vestibule is articulated by two Egyptian columns, with elaborate flower-capitals, which bear a cornice adorned with the winged solar disk and with falcons; above this is the flat arch of the pediment. Inside, in deep niches to the right and left, are Statues of the deceased and his wife in Egyptian dress, carved in white limestone. The door in the rear wall of the vestibule is surmounted by the winged sun's disk and a Uraeus frieze. To the right and left, on pedestals, are two large serpents with the Egyptian double crown, the caduceus of Hermes, and the thyrsus of Dionysos. Above are shields with heads of Medusa.[88]

Exciting descriptions and the sites they touted were meant to keep visitors in Alexandria for just a few more hours and were often successful. The artefacts from the complex, which had been taken to the Greco-Roman Museum, had accomplished this, too. After Botti's death, Evaristo Breccia held the position of director of the museum – and of archaeology in Alexandria – until 1932. He was the person in charge of continuing the excavations at the catacombs, providing accessibility to the site for scholars and the public. This archaeological work clearly had a direct impact on the visitor experience.

Although there had been significant changes in the town itself and in the archaeology of Alexandria from 1902 to 1914, the corresponding editions of the guidebooks demonstrated little difference in the discussion of Alexandria's tourism industry. Alexandria remained a transitional space between the European/Western comfort zone and the Egyptian/Eastern newness experienced by most travellers in this period. Travel writer Douglas Sladen even told readers who carried his 1911 *Queer Things about Egypt* in their travels: 'Alexandria is an Italian city ... [e]ven its ruins are Roman ... A few of the classical ruins are showing, most of the rest lie undisturbed under the mounds between Alexandria and Aboukir. Another Roma may await their investigator. Alexandria consists therefore of history and unhistorical buildings.'[89] Despite this destructive Western view, Breccia and others continued excavating important sites, and the remains that were not lost to the rapid expansion of the city were housed in the Greco-Roman Museum.

One major development appeared on the tramline to the exclusive resort suburb of Ramleh in the form of a newly discovered Ptolemaic necropolis. This necropolis might only pique the interest of the rich, who were on their way to Ramleh to rent a villa, hunt, or go to the beach.[90] As before, tourists were directed to see the director's office at the museum to get a permit to visit the new site. The main difference between the 1902, the 1914, and then the 1929 editions of Baedeker's *Egypt* was the notable urban expansion of the city itself. Expansion of the modern city was likely at the expense of ancient sites, as much of the city was being built on top of or with the Roman rubble. To attempt to combat the loss of important information, by the early decades of the twentieth century, excavations had revealed a number of important sites and had changed scientific and public ideas about Alexandrian remains. In turn, these developments made guidebooks more useful and descriptive when it came to escorting the traveller through the 'new' Egyptian city. By 1914, more tourists were stopping in Alexandria and staying for relatively extended periods of two days or more (Figure 1.4). The main sites of Pompey's Pillar and the Serapeum, the museum, and the Catacombs were well known and were on the beaten path for visitors, a far cry from the Pillar being the only monument in town.

The First World War and after

By the time of the final Baedeker's guide for Egypt in 1929, fifteen years and the First World War had irreversibly changed the city. The British declared war on the Ottoman Empire in 1914 and by the same stroke they made Egypt an official protectorate, ending the Khedivate of Egypt and installing Hussein Kamel as the Sultan. As far as Egypt was concerned, the War was most keenly felt in Alexandria, which continued its role as a

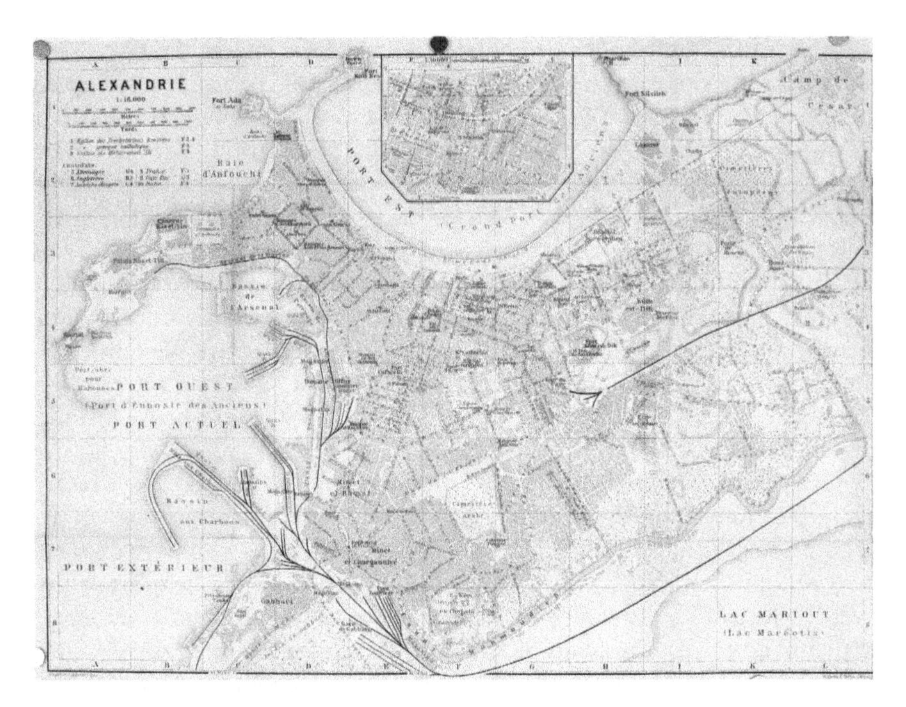

Figure 1.4 Baedeker's map of Alexandria, 1914.

transition space for Europeans and a new kind of traveller – the war weary. Much of the city was taken over by the Allies for administrative work, troop staging, and convalescent care. The harbour was the landing place for troops training for fighting in the Balkans and across North Africa. Divisions of troops landed in Alexandria to prepare for months of training in both Camp Mex near Alexandria and Camp Mena on the Giza Plateau.[91] The chaplain of the twenty-ninth division, Reverend O. Creighton, recalled his time preparing for and going to Gallipoli. In this preparation, he said, for operations, 'Alexandria, of course, was *the* base.'[92] Throughout the war, wounded troops were sent to Alexandria to convalesce away from the guns of the various fronts.

Egyptology and tourism largely came to a halt all over Egypt and British Khaki took over hotels throughout the country, but there was some work happening further south in Luxor. Military members took advantage of the sights, sounds, and experience of Egypt when they had some time off. The years between 1914 and 1918 were formative years for author E. M. Forster working as a Red Cross volunteer in Alexandria, falling in love with both the city and his trolley driver. He loved the city so much, he researched and wrote his own guide to it, *Alexandria: A History and A Guide*, published in 1922; he loved Mohammed el Adl so much that he dedicated the essays in *Pharos and Pharillon* to him the following year.[93] Through these works, he brought the city to life for readers all over the world.

After the War, tourists were slow to return, but by 1929, there were a number of hotels and pensions in the city catering to all budgets. Many of these were considered resort hotels, with amenities for long-term stays for full-time residents of Egypt to escape the heat of Cairo and points further south over the summer. Into the 1930s, casino hotels and other luxuries would be built and enjoyed by thousands of people. There were restaurants, cafés, and clubs. Even Groppi, the famous Swiss pastry shop based in Cairo, had a shop in Alexandria to cater to visitors arriving in the harbour. The new aerodrome in Alexandria was under construction by 1929, soon to bring quicker air travel services to people going to India, as well as those coming to Egypt, from London.[94] In the hopes of keeping visitors for a few days, the sites of Pompey's Pillar and Kom el-Shukafa were cleaned up after the War and made more tourist-friendly. Visitors were now advised to spend two full days in Alexandria, with the first day given to the ancient sites and the museum and the second day to seeing the modern city.[95] Each site's description remained largely the same in the guidebooks, with Baedeker's *Egypt* even keeping the same map from 1902. The museum had grown its collections and footprint in the city. Baedeker still provided a map and detailed guide, and noted that visitors were now allowed a 'hand camera'.[96] A highlight of room six, which was really more of a corridor, was the 'Lifesize figure of Apis in granite, found in the Serapeum …'.[97] Archaeologists in Alexandria found the Serapeum – the remains of the temple that housed the black Apis Bull statue of Ptolemy I's god – increasingly important, and had recently discovered the Necropolis of Anfushi, on the former island of Pharos.

Conclusions

The year 1929 would be Baedeker's last Egyptian guide; Cook's last guide was also in 1929. Baedeker's was a German company and, due to a failing post-War economy and political system, they could not continue the publication schedule.[98] Other guides were becoming equally handy and ubiquitous, such as Murray's or MacMillan's, but the numerous travelogues on the market still left the city out. Approaching the middle of the twentieth century, there did not seem to be much more to say about Alexandria than there had been fifty years earlier. There were new archaeological sites to draw the visitor's attention and more people living there, but Alexandria remained a transitional space for travellers. They used the city as a place to arrive, gather their belongings, and move into the 'real' Egypt, the Eastern Egypt that began in Cairo. One author wrote this of European tourists as late as 1909:

> They arrive in the Western Harbour, land, and rush off to see Cairo, where there is more sunshine and less rain, also more to see and less commercial talk. They do not remain in Alexandria to study Graeco-Egyptian art; and after a day spent in Chérif Pasha Street travel south as quickly as they can. Owing to this, Alexandria is dull in winter – the tourist season proper for Egypt.[99]

Clearly, the city was not at all dull. There were must-see historical sites and comfortable hotels for Europeanised excursions from Cairo in the summers. The city, which had been a village of 5000 people in the 1820s, by 1929 had turned into a vibrant city of over half a million people discovering its own past. As excavations continued, archaeology clearly changed *what* tourists saw and *how* they saw it. The sites and their material remains changed how long travellers stayed in Alexandria. Pompey's Pillar went from being the 'only site' in town to being the centre of a massive complex of tombs, temples, and artefacts. The Serapeum and Kom es-Shukafa were windows into a dynamic history of life in Alexandria, each with their own stories to tell. The Greco-Roman Egyptian centre of the world was, and still is, the Greco-Roman Museum. Alexandria would never be the 'exotic' locale sought by many Europeans and Americans, but by the 1930s it was, as Forster wrote, a city with 'the grandeur and the gentleness of the past behind a modern façade'.[100]

Alexandria was a significant place because it was the first entry into Egypt for most travellers around the turn of the twentieth century. It also became a significant place for archaeological tourism. The elegant coastal hotels were well appointed and gave tourists a taste of what was to come, as well as an entrée into Egyptian culture. Alexandria was Europeanised, but still Other enough to allow tourists a small cultural shift without a major culture shock as they moved into the East. The archaeologists present throughout this book moved through the city as quickly as tourists did, despite the growing interest in the city's ancient past. Because the Greco-Roman period is not seen as part of Egyptology strictly, but is seen as separate and after the Pharaonic periods (c. 3100–332 BCE), many archaeologists were interested in the sites, but as tourists and not in a scholarly way. The ways in which guidebooks sited and curated visitors' views of Alexandria as the transitional part-European, part-Egyptian space, would have a long-lasting impact on the study of the city's history. Alexandria has been placed outside of both the pharaonic past and the Greco-Roman past, and therefore the city has been left to its own field of study. It was not until later in the twentieth century that Alexandria became a significant Egyptological site for Egyptologists – and it remains unique.

The train to Cairo awaited these tourists and Egyptologists – many on tight schedules and budgets. Archaeologists would have sat in the second-class cars: not such a low class that they interacted with poor Egyptians, but not so high a class that they could not afford food for the day. As Baedeker's *Egypt* and Murray's *Handbook* told them to, many visitors tried to sit on the right side of the car, so they might see the outlines of the pyramids against the sky on the way into Cairo. The 130-mile trip could take between three and seven dusty hours.

Notes

1 Elizabeth Peters, *Crocodile on the sandbank* (New York: Dodd, Mead, 1975), 29. Egyptologist, and author of the Amelia Peabody mystery series, Elizabeth Peters (Barbara Mertz) had her fictional archaeologist write this in her 1885 journal.
2 Sometimes travellers came over from Suez (in Eastern Egypt), but for the most part this is the path they took.
3 This is not true for everyone, but Alexandria saw and sees a fraction of the tourists that Cairo does.
4 David Hogarth to Emily Paterson, 14 September 1895, EES.III.k.127.
5 Jean-Yves Empereur, *Alexandria rediscovered* (London: British Museum Press, 1998), 16–33.
6 See James Duncan, 'Dis-Orientation: On the shock of the familiar in a far-away place', in *Writes of passage*, eds Duncan and Gregory, 151–63.
7 These attitudes have been covered extensively. See, for example, classics such as Edward Said, *Orientalism* (New York: Vintage Books, 1979); Mitchell, *Colonising Egypt*; Mitchell, *Rule of Experts*. See also John Mackenzie, *Orientalism: History, theory and the arts* (Manchester: Manchester University Press, 1995); Derek Gregory, 'Emperors of the gaze: Photographic practices and productions of space in Egypt, 1839–1914', in *Picturing place: Photography and the geographical imagination*, eds Joan M. Schwartz and James R. Ryan (London: I. B. Tauris, 2003).
8 Baedeker's *Egypt* (1885).
9 Evaristo Breccia, *Alexandrea ad Aegyptum: A guide to the ancient and modern town and to its Graeco-Roman Museum* (Bergamo: Instituto Italiano d'Arti Grafiche, 1922), 9.
10 Jason Thompson, *Wonderful things: A history of Egyptology, Vol 2: The golden age: 1881–1914* (Cairo: AUC Press, 2015), 83–99.
11 Empereur, *Alexandria rediscovered.*
12 See Díaz-Andreu, *A history of archaeological tourism.*
13 See Amanda Vickery, *The gentleman's daughter: Women's lives in Georgian England* (New Haven: Yale University Press, 1998).

14 Victoria Peel and Anders Sørensen, *Exploring the use and impact of travel guidebooks* (Toronto: Channel View Publications, 2016).
15 Noted also by Gregory, 'Scripting Egypt', 120.
16 For example, Wilkinson's *The Nile* takes readers on a journey from the source of the Nile to Cairo, where the book abruptly ends.
17 Forster, *Alexandria*, xv.
18 Morsi Saad El-Din, 'Introduction', in *Alexandria: the site & the history*, ed. Gareth L. Steen (New York: NYU Press, 1993), 12.
19 Donald Malcolm Reid, 'The Urabi revolution and the British conquest, 1879–1882', in *The Cambridge history of Egypt (Volume 2)*, ed. M. W. Daly (Cambridge: Cambridge University Press, 1999), 217–38.
20 Baedeker's *Egypt* (1885), 218; Baedeker's *Egypt* (1895), 14.
21 Abdel/Azim Ramadan, 'Alexandria: French expedition to the modern age', in *Alexandria: the site & the history*, ed. Steen, 115–19; Baedeker's *Egypt* (1895), cxxxi.
22 Elisabetta Marino, 'Three British women travellers in Egypt: Sophia Lane Poole, Lucie Duff Gordon, and Emmeline Lott', in *The legacy of the grand tour: New essays on travel, literature, and culture*, ed. Lisa Coletta (Lanham: Rowman & Littlefield, 2015), 51–70.
23 Humphreys, *Grand hotels*, 23.
24 Amelia Edwards, *A thousand miles up the Nile* (London: Longmans, Green, 1877), 1.
25 *Ibid.*, 11.
26 Baedeker's *Egypt* (1885), 206.
27 J. Lane MSS 1, 27 November 1873, Journal 1 first half, Jenny Lane Collection, Griffith Institute of Egyptology, Oxford University. Boats would regularly be quarantined in Alexandria, often for forty-eight hours, but sometimes as long as two weeks, in an effort to stop the spread of infectious diseases.
28 Empereur, *Alexandria rediscovered*, 24.
29 David George Hogarth and Edward Fredrick Benson, 'Report on prospects of research in Alexandria: with note on excavations in Alexandrian cemeteries', *Archaeological Report (Egypt Exploration Fund) 1894–1895* (London: Macmillan, 1895). 1–33; E. A. Wallis Budge, *The Nile: Notes for travellers in Egypt*, 9th edn (London: Thomas Cook & Son, 1905), 356.
30 Hogarth and Benson, 'Report on prospects of research in Alexandria', 4; Empereur, *Alexandria rediscovered*, 24.
31 Along with Baedeker's books, see also *A handbook for travellers in Lower and Upper Egypt* (London: John Murray, 1888).
32 Baedeker's *Egypt* (1885), 203.
33 *Ibid.*
34 For more on these obelisks, see Bob Brier, *Cleopatra's Needles: The lost obelisks of Egypt* (London: Bloomsbury Egyptology, 2016).
35 Baedeker's *Egypt* (1885), 218.
36 Hilda Petrie, 7 December 1897, in Drower, ed., *Letters from the desert*, 118.
37 *Ibid.*

38 Baedeker's *Egypt* (1908), 7. Clark's was run by Frank C. Clark out of New York. He 'provided relatively comfortable travel under generally primitive conditions'. He chartered cruises through the Holy Land – including Egypt – from 1895–1925 (Ruth Kark, *American consuls in the Holy Land, 1832–1914* [Jerusalem: The Hebrew University, 1994], 235).
39 Maggie Benson to Mary Benson, 4 January 1894, in Arthur C. Benson, ed., *Life and letters of Maggie Benson* (London: J. Murray, 1917), 163.
40 Baedeker's *Egypt* (1885), 12.
41 Baedeker's *Egypt* (1908), xxiv.
42 Humphreys, *Grand hotels*, 24–5.
43 Baedeker's *Egypt* (1908), 7.
44 Humphreys, *Grand hotels*, 25–37.
45 Baedeker's *Egypt* (1902), xviii.
46 Drower, ed., *Letters from the desert*, 32.
47 Baedeker's *Egypt* (1878), 203. See Table 0.1 for conversion rates: 12.5 francs was about 10s in 1878. This roughly corresponds to the purchasing power of £35 or $53 today.
48 Playfair, *Handbook to the Mediterranean*, 50.
49 Hilda Petrie, 7 December 1897, in Drower, ed., *Letters from the desert*, 119.
50 *Ibid.*
51 James H. Breasted Correspondence, Oriental Institute, University of Chicago, Chicago, Illinois.
52 JHB Papers Box 4, Sept-Dec 1894, 28/29 Oct 1894. JHB Correspondence.
53 Foy D. Scalf, *The research archives of the Oriental Institute: Introduction and guide* (Chicago: Oriental Institute, 2013), 4. https://oi.uchicago.edu/sites/oi.uchicago.edu/files/uploads/shared/docs/research_archives_introduction&guide.pdf (accessed 12 December 2019).
54 JHB Papers Box 4, Sept–Dec 1894, 4 Nov 1894. JHB Correspondence.
55 *Ibid.*
56 *Ibid.*
57 Humphreys, *Grand hotels*, 25.
58 *Ibid.*, 27. In 1867, Mark Twain thought Alexandria too European a city to be novel and tired of it quickly (*The innocents abroad, or The new Pilgrims' Progress*, Vol. II [London: Harper & Brothers Publishers 1899], 400).
59 JHB Papers Box 4, Sept–Dec 1894, 4 Nov 1894. JHB Correspondence.
60 *Ibid.*
61 *Ibid.*
62 *Ibid.*
63 Baedeker's *Egypt* (1885), 222.
64 Baedeker's *Egypt* (1898), 5.
65 *Ibid.*
66 *Ibid.*, 6.
67 *Ibid.*
68 Baedeker's *Egypt* (1895), 17, 20.
69 *Ibid.*, 16–20.

70 Baedeker's *Egypt* (1902), 15–17.
71 *Ibid.*
72 *Ibid.*
73 *Ibid.*, 17.
74 Giuseppe Botti, ed. *Bulletin de la Société Archéologique d'Alexandrie*, 1 (Alexandria, 1898).
75 Giuseppe Botti, ed. *Plan de la ville d'Alexandrie à l'époque ptolémaique Monuments et localités de l'ancienne Alexandrie; Mémoire présenté à la société archéologique* (Alexandria, 1898)
76 Botti, ed. *Bulletin de la Société Archéologique d'Alexandrie*, 53–4.
77 See Table 0.1 for conversion rate. In 1908, 3 piastres would have been about 7d. This corresponds to the purchasing power of about £2.42 or $3.35 today.
78 Baedeker's *Egypt* (1908), 14.
79 *Ibid.*
80 *Ibid.*
81 *Ibid.*, 15; Baedeker's *Egypt* (1929), 18.
82 Baedeker's *Egypt* (1902), 13.
83 *Ibid.*
84 *Ibid.*
85 *Proceedings of the Society of Antiquaries in London* (1905), 157.
86 Baedeker's *Egypt* (1902), 14.
87 Baedeker's *Egypt* (1914), 17.
88 *Ibid.*, 18.
89 Douglas Sladen, *Queer things about Egypt* (Philadelphia: J. B. Lippincott Co., 1911), 154–5.
90 Baedeker's *Egypt* (1914), 26.
91 Rev. O. Creighton, *With the twenty-ninth division in Gallipoli: A chaplain's experiences* (London: Longman's, Green & Co., 1916), 26.
92 *Ibid.*
93 Forster, *Alexandria*; E. M. Forster, *Pharos and Pharillon* (New York: Alfred A. Knopf, 1923).
94 Baedeker's *Egypt* (1929), 9. For more on air travel, especially from Britain, see Gordon Pirie, 'Incidental tourism: British Imperial air travel in the 1930s', *Journal of Tourism History* 1:1 (2009): 49–66.
95 Baedeker's *Egypt* (1929), 9.
96 *Ibid.*, 22.
97 *Ibid.*, 23.
98 Wind, 'Profiles: The House of Baedeker', 92.
99 Arnold Wright, *Twentieth century impressions of Egypt: Its history, people, commerce, industries, and resources* (London: Lloyd's Greater Britain Publishing Company, 1909), 409, quoted in Haag, *Alexandria*, 16.
100 Forster, *Alexandria*, ii.

2

Cairo: the city and tourist victorious

Although travellers would have been in Egypt for at least a few hours, if not a couple of days, many marked their official entry into Egypt as the time when they finally pulled into the Cairo train station. Cairo, not Alexandria, was the first place many travellers wrote about when beginning their journals or memoirs. They felt they had finally arrived in the country when they could see the pyramids out their train window, the minarets of the city, and experience what they considered to be the truly Egyptian city of Cairo.

The approach to Cairo on the ground was and is vastly different than the perspective travellers get when approaching by air today. Today, depending on pollution levels and time of day, the airplane's shadow may be visible on the desert floor below, as well as the Nile and possibly the pyramids. Before the first rail line on the continent of Africa was completed in 1856 to bring travellers directly to Cairo from Alexandria, travellers would have to get on a boat that would snake through Alexandrian canals to the Nile, then up the river. By the late nineteenth century, travellers entered Cairo on a train, likely making a short stop at the Delta town of Tanta. They went through a number of smaller towns on the way, and in the three to seven hours the journey took them, they got more and more tired, hungry, and dusty.

Getting nearer the city, if it were still daytime, travellers might read their guidebook to know what they should look for as they anticipated their arrival. Beginning in Baedeker's 1878 edition, and continuing virtually unchanged over the next fifty years, the opening description told readers that, about twenty miles outside of Cairo

> the outlines of the pyramids begin to loom in the distance towards the S.W., and [ten miles further] near … Ḳalyûb these stupendous structures become distinctly visible … The Libyan [mountain] chain becomes more distinctly visible, and we also observe the Moḳaṭṭam range with the Citadel, and the Mosque of Moḥammad 'Ali with its slender minarets. The scenery now becomes more pleasing. The fields are enlivened with numerous trees, and gardens and villas come in sight.[1]

The description is echoed in Mary Brodrick's revised 1900 edition of Murray's *Handbook*, and a number of other published memoirs and private journals.[2] In 1886, Wallis Budge described his experience of taking the train from Alexandria into Cairo: '... when nearing Cairo I caught a glimpse of the two larger of the Pyramids of Gizah, standing out like a pair of twin breasts against the red light of the western sun'.[3] He was surely not the last man to use this comparison. In 1894, James and Frances Breasted came into Cairo on the train from Alexandria, riding second class. James wrote home to his parents that he 'discovered the Pyramids on the horizon with such a shout that our native fellow voyagers started from their seats in astonishment'.[4] Just three years after the Breasteds, newlyweds Hilda and Flinders Petrie arrived for their first season together in Egypt. She described what she thought of the pyramids when she first saw them as 'even mightier and more interesting than I expected'.[5]

A few more miles on the train past the view of the pyramids, travellers came into the centre of Cairo; 131 miles after leaving Alexandria, they disembarked at the busy Principal train station, situated to the northwest of central Cairo. Surrounded by tourists, archaeologists, artists, government officials, and locals, travellers would try to get to their final destination. They were often aided by hotel porters and donkey boys, and hindered by those trying to sell trinkets or fake artefacts. The guidebooks had already warned visitors in advance to book a hotel room before arrival in the city. Murray's *Handbook* informed new visitors that 'Omnibuses, carriages, and donkeys await the traveller. If the traveller has no one to meet him, he had better put himself into the hands of the commissionaire of the hotel to which he intends to go'.[6] From the train to their chosen (or first available) hotel and the activities they participated in all over the city, no two visitors experienced the same place. This chapter is, like the rest of this book, partly a history of archaeological tourism in Cairo. The sights, sounds, smells, and experiences in Cairo will be central to understanding all visitors to the city around the turn of the twentieth century. Before we can discuss how archaeologists used the city and its hotels as central scientific meeting spaces, we must first understand what they did as, and with, travellers.

This chapter will move into an examination of the activities of archaeologists and other visitors to this ancient city on the Nile, in order to piece together the importance of the city, its hotels, their inhabitants, and the transient visitors to the practice of archaeology during this period. The activities in Cairo will show that it was through this impermanence that the cognitive topographies and crucial informal networks rose and fell. Even though Cairo was a relatively brief stopover compared to the rest of the archaeological season, it was a productive stop for the archaeologists whom I will discuss in the following pages. Emma Andrews, James Breasted, Howard

Carter, Theodore Davis, Harold Jones, Flinders Petrie, Charles Wilbour, and others would meet, decide upon work, and get ready for the season. Their tasks included visiting the Egyptian Museum to get permissions from the Department of Antiquities, waiting for their crews to arrive in country, hiring new crew members, taking their families and excavation crews to see the sights, shopping for supplies for their seasons in the field, and talking to other archaeologists and patrons in the city. These activities were crucial in positioning each of our scientists as productive scholars for the season in the field, and beyond, so being in Cairo for at least a few weeks was a necessary part of most Egyptologists' journeys.

Hotels in Cairo were therefore different from the hotels in Alexandria, which were either useful for lunch or an overnight stay, and the hotels in Luxor, which were longer-term work/public and domestic/private spaces. In Cairo, hotels were certainly nodes of activity for our archaeologists, and foundations from which they performed a number of professional and personal tasks. First, they were both destinations to be enjoyed and staging grounds for further travel. Archaeologists would wait in Cairo for other people travelling or working with them for the season. This included family and crew members from their universities or professional societies. Related to the first task was the job of hiring more crew once they arrived in Egypt. Often, hotels were used by new and aspiring Egyptologists, artists, and excavators as places to ask for and get jobs. They also allowed both new and seasoned archaeologists to establish professional networks for new and continued patronage, study, and work. Third, hotels were used as informal meeting spaces where archaeologists discussed new professional ideas and projects with each other. These spaces were central to Egyptologists and the creation and sharing of knowledge. While hotels could be defined as truth spots in a number of ways, that is places in which ideas gain credibility and momentum, and which are central to the increased believability of those ideas, understanding how ideas flowed through the initial meeting in a hotel, to the excavation, and then into various publications, is very hard to trace.[7] We do know that hotels allowed archaeologists to share information in a much less formal way than they did in institutions like museums or on site, and that they came to be places where travellers could expect to find their colleagues and friends for professional and personal chats.

Finally, individually and together, hotels in Egypt were European spaces in a foreign land. They were spaces deeply imbued with colonial meaning in which Western Egyptologists created knowledge about Egypt's past. Egyptology was and is a colonial discipline, excluding most Egyptians from the practice until the 1920s.[8] Western visitors viewed Egyptians as the Other, a group to see as picturesque and exotic but not to engage with. This was clear nowhere more than in Cairo, where hotels were built on raised platforms that actually lifted Western visitors off the pavement and elevated them at least a few feet from the street life below. Westerners could view Egyptian street life from this lofty vantage point, watching the masses streaming past, but at the same time feeling as though they were a part of the excitement themselves. With Egyptians not allowed in most Western hotels unless they worked there or had special permission to enter, it may not be surprising that they were also excluded from studying their country's own history.[9] There were a few exceptions to this rule, especially in the career and work of Ahmed Kamal, but he was often excluded from educational and scientific institutions because of the hostile colonial environment.[10] By physically excluding Egyptians not only from educational institutions, but also from the critical informal hotel space, Egyptologists excluded Egyptian voices from the discipline of Egyptology. Using hotel terraces, dining rooms, and private suites as meeting places, these Western Egyptologist sate and talked with one another and with other visitors, thus lending authority to their plans and ideas.

In this chapter I present the city of Cairo; al-Qahira; Fustat; Memphis. It was the nineteenth-century jewel of the Nile and the ancient Near East. Before it is possible to talk about the activities of travellers within and outside of hotels, it is important to get to know the city – its layout, its streets, its people, the sounds, sights, smells, and feel. I will introduce and describe hotels and centralise these liminal spaces as the places where table talk on the terrace reigned supreme, thus making them each a node of a cognitive topography that became an ephemeral network of people, sites, and ideas.

Arriving in the city

The city of Cairo, Al-Qahira, was founded in 969 CE by the Fatimid dynasty, but the city's history can be traced back to the ancient Egyptian capital of Memphis, founded over 5000 years ago, about twelve miles south of present-day Cairo on the banks of the Nile. The area has been an important settlement for millennia. There are remains of Roman forts in what is now the middle of the city, along the Nile River. All of this is not to mention the 5000-year-old pyramids on the Giza Plateau, which, on a rare clear morning today, you may be able to see from the vantage point of the Citadel. There has been a lot written about the fabled ancient history of Cairo and its inhabitants, so I will not repeat what others have done better and more fully than I can do here.[11]

Cairo has been the capital of Egypt since 1168, when it moved from Fustat (now in Old Cairo). The city's location was crucial: at the entrance to the Delta, the pyramid field lay to the West, stretching south up the Nile. On the way to Suez, before the Canal had been built, Cairo had been a way-station on the old overland route. It was the centre of culture, society, and travel in Egypt and much of the colonised world. It held, and holds, foreign consulates and embassies; it houses diplomats, businessmen and women, and the seat of government. It was a central staging ground for pilgrims' journeys to Mecca and Medina, centuries before it was the centre for Europeans travelling up the Nile. Even though Alexandria was a major port, providing access to the Nile from the Mediterranean, most goods were bought and sold in Cairo. Often tourists would stay the whole 'season' – October to April – in Cairo (Figure 2.1). In these months it was not too hot for Europeans to visit and work. Many used the city as a starting point for further travel up the Nile. For others it was a brief stopover on their journeys to other parts of the Middle East or India. Regardless of their purpose for being in Cairo, visitors met their groups, saw the sights, rented boats, and bought supplies for long journeys or seasons in the desert.

The Principal Station (from 1907 the Central Station; currently Ramesses Station), the station that served all Alexandria trains incoming and outgoing, was on the northwest side of the Ismaliyeh canal, a half mile from the end of the Muski.[12] In 1885, guidebooks told visitors that upon their arrival, 'hotel commissionaires with their omnibuses or carriages await the arrival of each train and take charge of luggage'.[13] By 1895, Baedeker's *Egypt* added to this list of greeters at the newly refurbished station 'Cook's and Gaze's agents' in reference to the European-based travel agents, Thomas Cook & Son and Henry Gaze, who would greet tourists on their first, or tenth, visit to Egypt.[14] The hotels in 1885 included the New Hotel, Shepheard's, Hotel du Nil, Hotel Royal, Hotel d'Orient, and Hotel d'Angleterre.[15] Most visitors needed to make sure they had telegrammed ahead for rooms because hotels could be full during the tourist season. Added to the list of hotels in 1895, among others, were the Continental, Gezirah Palace, and Mena House Hotel; not on the list anymore was the Hotel d'Orient.[16]

Baedeker's informed travellers that planning ahead for sightseeing would also suit them well:

> By carefully preparing a plan beforehand, and starting early every morning, the traveller may succeed in visiting all the chief objects of interest in six days, but it need hardly be said that a satisfactory insight into Oriental life can not [sic] be obtained without a stay of several weeks.[17]

Unlike travel advice for visiting Alexandria, which expanded the tourists' experience over the years from a half day to around two-and-a-half days thanks to archaeological discoveries, the basic tourist advice for Cairo did not change for the next thirty years.[18] Several days to several weeks were

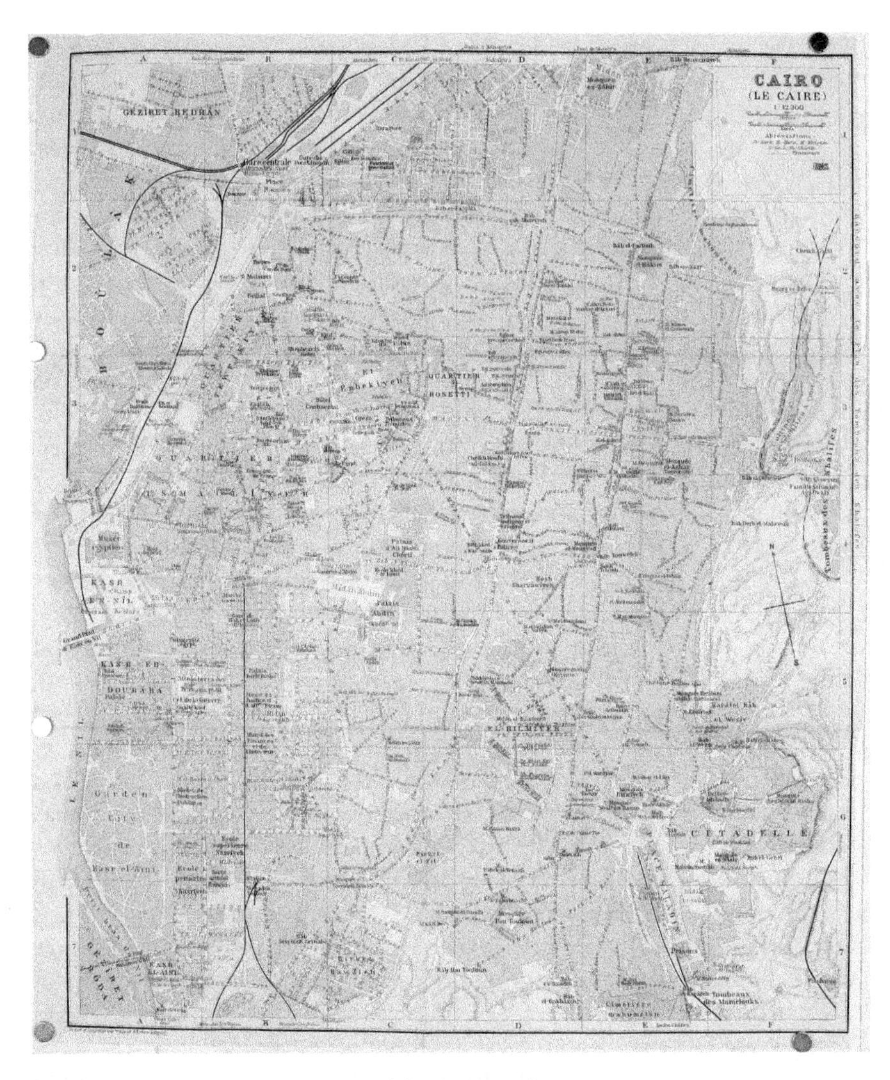

Figure 2.1 Baedeker's map of Cairo, 1914.

suggested for a full experience and hotels were expanding to accommodate the influx of tourists. Thomas Cook had made Egypt more accessible to Western tourists from the time of his first guided tour there in 1869, so prices were becoming more reasonable for many people.[19] By 1914, there were plenty more hotels – most of them fitted with electricity – to accommodate the influx of guests, and travellers could be sure that the porters from each were ready and waiting for them at the station.[20] The Semiramis, Continental, Gezirah Palace, and Mena House were a few new hotels that visitors could choose from in 1914. There were Pensions, cheaper lodgings with full room and board, for those with tight purses; for those who wanted more luxury and privacy, there were private apartments and houseboats. Of course, Cook's and other tourist agents and their myriad options abounded.

For diversion while in the city, in addition to shops, theatres, clubs, and churches, Baedeker's and other guidebooks led travellers around Cairo much the same as they did in Alexandria. There were step-by-step instructions throughout the guide for visiting different places and detailed, prescribed travel itineraries. 'Energetic travellers' were given, at minimum, six days, but it was suggested that it would take eight days to see Cairo and its surrounding sights properly. If they made 'free use of cabs', instead of slower donkeys, travellers would likely be able to accomplish much. A detailed itinerary of these days is given in each available guide, whether it was Murray's, Baedeker's, or Cook's, including the route and conveyance to take. Table 2.1 displays a general summary and comparison between the common 1895 and 1914 itineraries here.

In 1895, on the fourth day, visitors to the Museum at Giza (the former location of the Egyptian Museum) were within a few miles of the pyramids,

Table 2.1 Cairo itineraries

	1895[a]	1914[b]
Day 1	Morning: tour Azbakeya Garden, then to the Muski and the bazaars Afternoon: tombs of the caliphs and the Citadel	Morning: tour Azbakeya Garden, then to the Muski and the bazaars Afternoon: tombs of the caliphs and the Citadel, mosque of Mohammed Ali
Day 2	Morning: Visiting mosques of Sultan Hasan, Ibn Tulun, El Muaiyad and El Ghuri Afternoon: Go out to Matariyeh and Heliopolis	Morning: Egyptian Museum (by this point just off what was then Midan Mariette Pasha, what is now Tahrir Square) Afternoon: Visiting mosques of Sultan Hasan, Ibn Tulun, Kait Bey
Day 3	Morning: More mosques like that of ibn Kaldun, and a visit to the Arabian Museum[c] Afternoon: View the dervishes on the Kasr al Aini and then cross the Nile bridge to visit Gezirah Island	Pyramids of Giza, 'which may be seen in the course of a forenoon, if necessary'.

Table 2.1 Cairo itineraries (Continued)

	1895[a]	1914[b]
Day 4	Morning: Museum of Giza (built here in 1891, after having been in Bulaq for 35 years) Afternoon: Mokattam Hills	Morning: El Azhar, Ghuri and Muaiyad mosques, then bazaars Afternoon: Go out to Matariyeh and Heliopolis
Day 5	Morning: El Azhar and Hasanein mosques, then bazaars Afternoon: Island of Roda and Old Cairo with the Coptic churches, possibly the Tombs of the Mamlukes	Morning: Second visit to the Egyptian Museum, then to the bazaars Mokattam hills and the monastery of the Dervishes
Day 6	Pyramids, which, the guidebook tells the visitor, 'may be seen in the course of a forenoon, if necessary', then a visit or return through Shubra in the afternoon	Morning: Arabian Museum Afternoon: Across the Kasr el Nil bridge to visit Gezirah and the Zoo
Day 7	Memphis and Sakkara pyramid fields	Memphis and Sakkara pyramid fields
Day 8	Barrage du Nil (on a steamer)	Morning: Mosques and Bab en Nasr Afternoon: Island of Roda and Old Cairo with the Coptic churches, possibly the Tombs of the Mamlukes
Day 9		Barrage du Nil
Day 10		Abu Roash or Abusir pyramid fields

[a] Baedeker's *Egypt* (1895), 33–4.
[b] Baedeker's *Egypt* (1914), 43–4. See also Brodrick, *Handbook for travellers*, 319–22.
[c] By 1914 the Arabian Museum was called the House of Arab Antiquities. In 1951, its current name, the Islamic Museum, was made official.

but it was suggested to save them for a different day. There were no ninth or tenth days included in the 1895 itinerary, but Baedeker's suggested revisiting a few important places, including the Museum at Giza, the Citadel, the Tombs of the Caliphs, Azbakeya, and the bazaars with street traffic (especially heavy on Thursdays when people shopped before the Muslim Holy Day on Friday).[21] Both itineraries suggested visiting the museums and bazaars many times, in order to really experience the feel of the city.

In 1901, the new Egyptian Museum was opened in what was then Midan Mariette Pasha, and is now Tahrir Square, on the banks of the Nile in the area known as the Bulaq. This new location made it easier for tourists staying in the city to visit the museum, because it was not out by the pyramids, which were ten to twelve miles from the city. The museum's change in location would make a drastic difference on the order of the proposed itinerary for visitors. Murray's *Handbook* in 1907 gave travellers two itineraries for Cairo – one lasting three days and one lasting nine. Both included the pyramids at Giza and Saqqara, the museum, and Old Cairo. The main difference between them was time spent at each place. The museum and allotted time, along with a number of archaeological discoveries, would be the main factors in the changing of the itinerary in 1914, just before the First World War. By 1916, Reynolds-Ball's *Cairo of To-Day* included many of the same sites as Murray's and Baedeker's, along with a significant section about Cairo as a health resort.[22] By 1929, itineraries had changed again, but largely to add more than one visit to a site if the visitor had time. Travellers in Cairo obviously had many choices for sights, sounds, smells, and tastes. While many of them would have followed the above general Baedeker's itinerary, others may have followed Murray's, Cook's, Usborne's, or other private memoirs as guides.[23] The order of viewing might be different, but the highlights were the same. No matter how long the visit, there were some central areas of importance for all visitors to the city, and to the archaeologists staying there. Included in this list were the Azbakeya Gardens, the Muski, the Bulaq area, and the Nile River, all within walkable proximity in the city. Outside the city, the Giza Plateau was and still is a necessary stop. I will detail each of these places and their importance to visitors in terms of archaeological tourism before going into detail about archaeologists' activities in and around these places.

Azbakeya Gardens

One historian argues that 'long before the British came you could draw a north and south line through Cairo, down the edge of the Ezbekiya Gardens, and you could say that almost everything to the east was the old medieval Arab city and everything to the west (except Bulaq) was the new European one …' (Figure 2.2).[24] For centuries much of the district known as Azbakeya, named after Azbek (Uzbek) min Tutukh, a Circassian emir and commander of the Sultan's armies who developed the area in the fifteenth century, had been a swampy area, holding the slowly drying remnants of the annual Nile inundation and piled with waste.[25] It was a pool of stagnant water that bred mosquitos, therefore mosquito-borne illnesses, and death. When Napoleon arrived in Cairo in the summer of 1798, he moved into the area

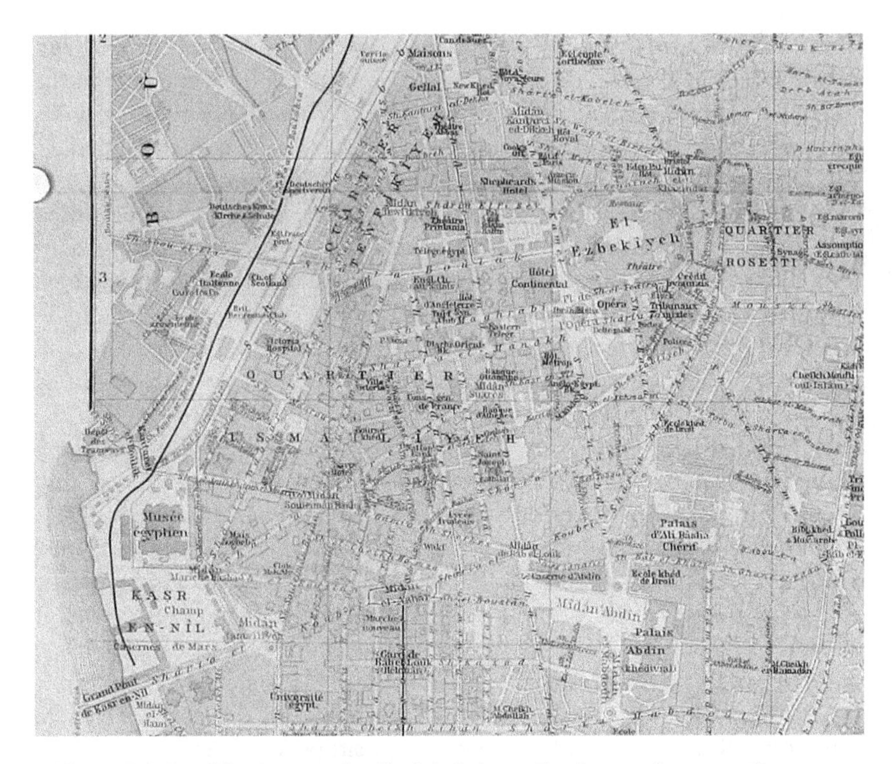

Figure 2.2 Baedeker's map, detail of Azbakeya Gardens and surrounding area, 1914.

and commandeered a 'brand-new palace which Mohammed Bey al Elfi had built at great expense but had not yet occupied'.[26] On this spot would be built the home of the first famous Shepheard's (British) Hotel, but not for another twenty or so years. European-style cafés and amenities began to open and thrive with Western businesses while the French military occupied Cairo, from 1798 to 1801. After the British defeated the French in the Battle of Alexandria in 1801, and with the signing of the Treaty of Paris ending the conflict in 1802, Egypt was effectively handed back to the Ottoman Empire. But European control never really left. The Gardens remained an area of European influence and a symbol of the stronghold of colonial power in the country. In the years before they were filled in with landfill and kept dry by drainage ditches, the inundation would turn the gardens into a lake on which boats would sail for leisure during the day and at night.[27] By the mid-nineteenth century, however, the lakes and stagnant pools of water had been mostly filled in, due to European engineering works. A canal had been built around the space, and European homes had sprung up in the area.[28]

For more than a century, from the early nineteenth century to the 1952 revolution that brought about the end of British occupation, these were the central gardens of the massive and growing city of Cairo. Here was the central place for Europeans, Americans, and Egyptians to meet and interact.[29] Egyptologists and tourists alike would have known the area well, calling the hotels in the area home for the length of their stay in Cairo. Thomas Cook, whose offices were just off the Gardens and next to Shepheard's Hotel from 1873, told readers of their 1897 *Tourist Guide*:

> [The] Esbekeeyah is the most important public place in Cairo, adjacent to several places or official buildings – the Opera House, the Palace, and the chief banks and hotels. There is a large garden in the centre, surrounded by a fine avenue, [with] alleys of trees radiating from the centre. Santi's restaurant is famous for *déjeuners* and dinners. During the evening the open-air cafes are well patronized, but European residents of the better class are seldom seen in the garden.[30]

Jenny Lane, who also enjoyed Alexandria while travelling with Amelia Edwards and Lucy Renshaw in 1873, delighted in her time in the Gardens listening to a band, but battling mosquitoes.[31] In 1907, Norma Lorimer, a British writer travelling on her own through Egypt, was particularly struck by the area. She wrote:

> However long I stay in Cairo, this quarter of the Ezbekiyeh will always maintain a paramount place in my vision of the city. It is so strange, so picturesque, so unceasingly amusing, so changingly changing, so simple and yet so depraved ... Here was the most delightful feast of colour and oriental happenings you could ever hope to find.[32]

During the nineteenth century, as the British and French gained more political and military control over Egypt, more and more tourists came to Cairo in hopes of finding this 'feast' of 'oriental happenings', and they flocked to the Gardens to witness it. They would stay at hotels bordering the Gardens to do so: Shepheard's Hotel (previously Hill's and the British Hotel), the (Grand) Continental, the Eden Palace, and the Hotel d'Orient. From the clean safety of the windows and terraces of these European hotels, visitors could watch the people below from a comfortable distance while not necessarily taking part.

The Gardens have changed in the decades since 1952, but the area has been a place of change for centuries. Where there is now pavement with a highway overpass as its roof, people used to stroll under shade trees to escape the bustle of street traffic. Hardly a trace remains of the once shady walks, luscious gardens, ponds and pools, kiosks serving teas and coffees, and gazebos with live music. The Gardens were central for archaeologists, as we will see, as the area had the best and healthiest breezes and most convenient places to stay, meet, and work.

The Muski

The gardens were next to what was known as the Muski, or the quarter of town where Egyptians lived, worked, and shopped. The Muski was also the name of the main street that ran through this quarter. It was so busy and crowded that some likened it to New York City and called it the 'Broadway of Cairo'.[33] Cook described it in 1897 as 'a fine street running from the Esbekeeyah through the very heart of the city. It forms the Frank Quarter, and is well provided with shops.'[34] There were antiquities shops and shops that sold almost any item visitors might need or want. The Muski was such an iconic part of the city that attempts to recreate it appeared in a number of World's Fairs and exhibitions throughout the nineteenth century and into the twentieth.[35] It was these recreations, Timothy Mitchell has argued, that turned Egypt into an object and a curiosity for Westerners.[36] While Mitchell focused on European recreations, an exhibit of the Muski was also built on the Midway at the 1893 World Columbian Exhibition in Chicago. In all of its colonial recreation, visitors could see the 'real' Cairo in the US with one major difference: it was safe, relatively clean, and close to home.[37] In fact, when James Breasted was still in Berlin, his parents, who lived in the Chicago suburb of Downer's Grove, went to the Exhibition to visit the Muski.[38] Back in Egypt, though, tourists continued to marvel at the picturesqueness of Cairo and to feel as though they were still watching an exhibit from the outside. As Mitchell noted, 'although they thought of themselves as moving from the pictures to the real thing, they went on trying ... to grasp the real thing as a picture. How could they do otherwise, since they took reality itself to be a picture?'[39] Those visitors who wished to experience the picture, and live in the real thing, could stay in the Hotel du Nil, deep in the quarter; a lot of early archaeologists did.

The Bulaq

Built by the Mamelukes, the area known as the Bulaq (Boulak, Bulak) has been a port on the Nile in Cairo since the fifteenth century. Money, spices, and other goods were passed through and traded in this port, making the area and its residents rich. Residents and area leaders built palaces and monuments to the port's success over the years. Over the next three centuries, not a lot changed in the layout of the area, but its fortunes came and went. By the start of the nineteenth century, it was a run-down area in need of refurbishment to be useful to both locals and conquering nations. Through European contact and financial resources, the area was rebuilt as an important hub of activity. Muhammad 'Ali turned the district into an industrial area, aimed at providing the building materials and other resources for modernising the city.[40] In 1882, the British army occupied the Kasr el-Nil barracks near the Bulaq (but close enough to still be called the Bulaq), until they were removed by revolution in 1952.

The first Egyptian Museum was established in the old post office in the Bulaq in 1858, but because of the regular flooding of the river every year, the museum's contents were in constant danger of being destroyed by the river, dampness, and frost. The objects were moved to a former palace in Giza in 1891. And in 1901, the new Egyptian Museum was completed in the Midan Mariette Pasha (now Tahrir Square). The new, centralised location made it easier for all visitors to the city to stop for a few different sights and activities, including the Kasr el Nil Bridge, and across it, the island of Gezirah. It was on the southern end of the Bulaq area, near the Nile, but not in danger of inundation and rendered fireproof. Until at least 1919, the museum received at least half of all excavation finds in addition to objects of cultural importance; after 1919 the museum received almost all of them.[41] It was here that archaeologists came after their excavations to split the season's finds with the Department of Antiquities. The museum was a centre of scientific work and networking for archaeologists as well as a rich source of their ideas, and a place for tourists and Egyptians to visit to see pieces of history they could not view anywhere else.

The Nile

The River. In Northern Africa one hardly even needs to say which one. It is the Nile, a source of life and death for Egyptians for millennia. Its branches, the White Nile and the Blue Nile, run from present-day Uganda and Ethiopia respectively. Joining together in what is now Sudan, the Nile flows north and makes its way through Eastern Egypt. For ancient Egyptians the river separated the East, the land of the rising Sun and new life every day, from the West, the land of the setting, dying sun, death and the vast, dry desert. All visitors to Egypt today recognise the importance of the river to survival in Egypt. Most visitors in the period covered here would have spent considerable time on or near the Nile, in a felucca, ferry, dahabeah, or steamer, in order to go from one side of the river to the other, traverse the entire country, or to travel among various sites and towns. I will detail this river journey in the next chapter.

The Giza Plateau and pyramids

Located ten to twelve miles west of the centre of Cairo is the middle of a vast pyramid field dating back to Egypt's Old Kingdom (c. 2600–2200 BCE). The entire field extends for twenty to thirty miles up the West Bank

Figure 2.3 View of the pyramids, c. 1890.

of the river. The Plateau itself holds the Sphinx, nine pyramids, three of which are recognised as being the biggest that survive and in the best shape (Figure 2.3). There are also a number of rock-cut tombs, temples, and much more. To the north and northwest of Giza is the site of Abu Roash. Even though Baedeker's and other guidebooks suggested the sites as a day to itself, not everyone made it there. To the south of Giza lie the cemeteries of ancient Memphis, including Saqqara, Abusir, Dahshur, and Meydum. Again, being about ten to twelve miles south of the city, these sites would occupy a full day of time, even if tourists took the relatively quick railway from Cairo.

From the time of Herodotus and Strabo, almost every visitor to Cairo would go out to the Giza Plateau to see the only surviving wonders of the ancient world, but many would have to bring their own mattresses and camp in the open desert. There were no real roads to the plateau until 1869, when the dirt path was paved for Empress Eugénie who arrived to celebrate the opening of the Suez Canal; the first bridge over the Nile from Cairo – El Gezirah Bridge, now the Kasr El Nil Bridge – was not built until 1872. Up to that point, the only way to cross the river was by ferry and then travellers had to make long detours to avoid all of the land under the Nile inundation.[42] Once on the west side of the river, travellers had to find a donkey to ride, not that this was difficult to do. As Baedeker's pointed out, travellers to Egypt had long known that donkeys 'afford the best and most rapid mode of locomotion in the narrow and crowded streets of Cairo, and they are to be met with, day and night, in every part of the town'.[43] They were especially useful on the uneven and flooded ground on the way to the plateau. Despite the convenience, Thackeray had remarked that riding a donkey was 'not a dignified occupation'.[44] In 1867, the American writer Mark Twain quipped about his donkey riding experience in Egypt: 'I believe I would rather ride a donkey than any beast in the world. He goes briskly, he puts on no airs, he is docile, though opinionated. Satan himself could not scare him, and he is convenient – very convenient.'[45]

With all the difficulty of getting to the Plateau in the early days, it was often necessary to spend the night out on the site. Sometimes people would camp in the desert, but Mr Hill of the British Hotel (later Shepheard's) would also provide bedframes in the rock-cut tombs. Visitors had to bring their own mattress and bedding.[46] While this may sound outrageous to twenty-first century travellers, in the nineteenth century, archaeologists and less well-to-do travellers were doing this regularly, especially Flinders Petrie, who preferred the tomb over a hotel.[47] Much of this changed when the Mena House Hotel opened on the Plateau in 1886 with eighty guest rooms, a marble staircase, reception rooms, a billiard room, a darkroom where photographers could develop their pictures, and a library, from which, hotelier Auguste Wild noted, 'it was possible to quietly observe the tourists laboring up the face of the Great Pyramid'.[48] In 1895, Baedeker's *Egypt* touted Mena House as 'an extensive establishment, with various "dépendances", swimming and other baths, stables, riding-course, etc. (physician in residence) ... special terms for invalids ...'.[49] By 1900 the hotel advertised in the English-language newspaper *The Egyptian Gazette* that they had 'Concerts on Sundays and Wednesdays, Golf, 18 holes, 2 good Croquet Lawns, Tennis Lawn, Horses, Donkeys, and Desert Carts on Hire. Shooting. Pure well water. Milk from own Dairy and medically-examined cows.'[50] Mena House worked to make sure that it was not only where wealthy Europeans could relax, but also a place where 'sick people, in search of purer air; and consumptive English maidens; and ancient English dames, a little worse for wear, who bring the rheumatisms for the treatment of the dry winds' could find comfort.[51] It was well known that the air at the pyramids, at that time far removed from the crowded city, was a much better option than Cairo for patients with lung diseases and other ailments.

More than a hotel, Mena House also ushered in a new and better way (to some) to see the pyramids. An advertisement in *The Egyptian Gazette* in November 1894 touted the four-horse coach called the 'Mena-Ra' which would leave from Thomas Cook's office near Shepheard's on the Azbakeya at 11:45 am and would reach the pyramids at 12:55 pm. 'An excellent lunch is provided at the Mena House Hotel for the price of 4s' and then the coach would return to Cairo by 5:45 pm.[52] This would give visitors time to reach the top of the Great Pyramid and venture about inside and do a bit more sightseeing on the Plateau. By 1895, donkeys were 'becoming less and less fashionable', with the advent of trams and cabs, but were 'still indispensable' for some activities.[53] They were cheap and mostly agreeable, and the donkey boys were known to 'possess a considerable fund of humour, and their good spirits react upon their donkeys'.[54] By the turn of the twentieth century, more expensive but quicker cabs were the preferable way to go through the city and to the Plateau, and by 1899 there was a tramway that connected the Plateau directly with Cairo, making the journey in less than an hour. In 1902, Baedeker's told visitors that, once they arrived and settled in Cairo, they might take either the tram or the road to the pyramids. The road, which followed the tram line, left the Museum of Giza on the West Bank of the Nile.

> It there quits the Nile and runs inland, skirting the park of the palace [of Giza]. The prison lies of the left of the road; and on the same side are the village of Gizeh … The road makes a curve, crosses the railway … and then leads straight towards the Pyramids which are still nearly 5 M. distant … The huge angular forms of the Pyramids now loom through the morning mist, and soon stand out in clear outlines with all the injuries they have sustained during the lapse of ages. A few hundred yards before the road begins to ascend, it is protected against the encroachments of the sand by a wall 5 ft. in height. On the right are the extensive buildings of the *Mena House Hotel* … to the left is the terminus of the *Electric Tramway*. The road winds up the steep N. slope of the plateau on which the Pyramids stand.[55]

Baedeker's reminded travellers that they should choose a 'fine and calm day' to travel out to the pyramids, as driving in windy weather would kick up the sand and that would be 'very unpleasant'.[56] The tram was a popular option, and the Thomas Cook's coaches helped place Mena House at the centre of tourist need and attention, even though it was far from the city.

Baedeker's and other guidebooks at the time encouraged visitors to climb the pyramids as well, and in fact to prioritise this activity if they had no time for anything else. Most visitors marvelled at the pyramids, as we saw in the opening to this chapter, and they climbed. Harriet Martineau remarked in 1856 'I had been assured that I should be disappointed in the first sight of the pyramids … [but] so far from being disappointed, I was filled with surprise and awe; and … I felt as if I had never before looked upon anything so new as those clear and vivid masses, with their sharp blue shadows, standing firm and alone on their expanse of sand'.[57] Another travelling British woman, Amelia Edwards, initially thought the pyramids were 'not impressive' when she saw them as she was arriving in Cairo by train, but once she arrived on the Plateau, she had changed her mind. She recorded in *A Thousand Miles Up the Nile* that the Great Pyramid 'shuts out the sky and the horizon. It shuts out all the other Pyramids. it shuts out everything but the sense of awe and wonder.'[58] She visited more than once and, like so many others, climbed to the top of the Great Pyramid.

On their honeymoon in 1894, Breasted wrote to his parents that he and Frances began making their way to the pyramids at 8 am, and fully described the trip to the Plateau once they crossed the river.

> It is a beautiful drive under 5 miles of shady acacia trees & 2 ½ miles more of hot sun and then I stood under the shadow of the oldest human monuments on earth. I will not attempt any description nor say anything of my own feelings as I stood on the summit and looked down and the mighty mass beneath looked off over the inundated valley to distant Cairo where the white domed mosques and tall minarettes glittered in the sun, while between lay many a fertile village, cool under the vivid green of drooping palms; on turning around only, one cast his eye far out over the sun red barren sands of the Sahara, a picture of death, over which the tremulous waves of heated air proclaimed the vast furnace it was. And we went inside too, & came out & ate our lunch on the base amid a staring group of turbaned Bedouin ... But perhaps I can tell you more of it next time.[59]

On that first trip, the Breasteds also visited Saqqara and Dashur to the south. In 1919, Breasted also flew above the pyramids in an open biplane, getting motion sick while up in the air. In 1897, the Petries climbed to the top of the Great Pyramid, and Hilda made sure to record that she did it 'without the usual tedious help', that is, without having two or three Egyptians to pull and push her up the steep blocks on the pyramid slope, likely because she did it without her skirts, but certainly in breeches.[60]

As far as we know, few archaeologists stayed out at Mena House for an extended period of time. For most of them, their work was in the city, eight or so miles away, and they preferred to be closer to those activities.[61] British artist and archaeologist Howard Carter stayed at the Mena House frequently, and while he was the Inspector of Lower Egypt, he visited the hotel regularly. He enjoyed the fresh, dry desert air and it helped to cure his frequent colds and other ailments. In 1905, when he was stationed in Lower Egypt at Saqqara, he wrote that he had gone to the hotel in January to 'get rid of a very bad cold I have taken for a second time this month'.[62] By this point guidebooks argued that the spa town of Helwan, just south of Cairo, or the long Nile voyage by private houseboat were the best ways to restore one's pulmonary health, but as Mena House advertised and many like

Carter found out, the Giza Plateau was far enough away from the polluted air of Cairo to be beneficial.[63] Mena House had Egypt's 'undeniably superior climate' but was equally 'undeniably expensive, and the fashionable society element is too obtrusive to make it desirable winter quarters for the invalid'.[64] Due to the wealthier clientele of the Mena House, Carter and others likely did not plan many meetings at this spot with other archaeologists or openly speak to tourists. The hotel was not mentioned by many archaeologists in their letters. Norman de Garis Davies wrote quite a few letters back to the Egypt Exploration Fund's London offices in the 1904–05 season on Mena House letterhead, but it seems that he wrote them from smaller hotels in or near Luxor.[65] Either he stayed at Mena House at some point and took some paper, or the smaller hotels had left over paper from other travellers.

By the start of the First World War, Mena House was welcoming guests from all over the world. They were one of many hotels who took Cook's hotel coupons, making it more convenient for visitors to stay there.[66] During the First World War, tourists largely left Egypt, so it became Camp Mena, a war-time training annex for the British and ANZAC troops. Mena House was a significant hotel in Cairo for a number of reasons, but mostly as a layover for visitors only interested in seeing the pyramids. For others, it was a place to rest and recover from respiratory or other illnesses. This hotel had less impact on the practice of Egyptology than the hotels in the city because of its peripheral setting to all the activities visitors and archaeologists needed to finish before they left for places further south. But it was still a central place for tourists and the important visit to the pyramids all visitors must make.

Building archaeologists in Cairo: hotels, careers, and the intellectual landscape

Like other travellers, archaeologists, on their first trip to Egypt, would stop in Cairo for at least a few days if not a few weeks. Their experiences at the start of their trips and the start of their careers in Egypt were much like those of tourists. They often preferred to see the sights before they got to work. Many times, they had brought colleagues, students, and sometimes family members with them who might be arriving in Egypt for the first time, so being a tourist for a few days was necessary. Charles Wilbour sometimes brought his wife and family with him on trips up the Nile on his dahabeah the *Seven Hathors*. James and Frances Breasted honeymooned in Egypt from 1894 to 1895 and brought their young family to Egypt for the first time in 1905; Breasted continued to return on his own and sometimes brought his family from that point on. Flinders and Hilda Petrie also honeymooned in

Egypt in 1897 and brought their students and family each year until 1935. Theodore Davis and Emma Andrews, who travelled through Egypt every year from 1889 through 1913, always had guests with them. Harold Jones, a first time visitor and hopeful Egyptologist in 1904, wrote dozens of letters home to his family and friends on his first trip to Egypt, forming a sort of travelogue for them. Howard Carter, Margaret Benson, and countless others would introduce visitors as they travelled to and through Cairo.

In this chapter, I have established some main spaces and sites in which both archaeologists and tourists would have spent a lot of time doing important preparation work. They would have been rubbing elbows at many of the sites in and around Cairo, sometimes participating in the same tour groups that would be organised at the hotels. On the other hand, archaeologists were in the country for work and they had tasks at hand other than touring. They were organising their own work groups, readying students, spouses, children, and guests for the long and arduous field season. They were busy gathering their supplies from shops in the Muski and along Frank and Kasr el-Nil streets, famous for carrying European goods. The Egyptian Museum was one of the main centres of work for archaeologists while they were in Cairo. In the museum they would see finds from the previous field season, ask for and often (but not always) receive permission to excavate, and, at the end of the season, split their finds before heading home for the summer.

Cairo was an easy home base for excavation work at Saqqara and Giza. Archaeologists excavating at those sites would not necessarily stay in the city, but they could easily come to visit, take a break for a day or two, and get supplies. New and hopeful archaeologists, who were more like tourists at the beginning of their careers, had to begin to set up their work so that they could become professionals. Many of them arrived with job prospects, connections they wanted to make, letters of introduction to powerful people and many (but by no means all) of them arrived with some important job skills like training in art or excavation, or knowing how to decipher hieroglyphs or speak Arabic. They wanted to meet archaeologists or patrons in the hotels so that they could get themselves a job to last the season, maybe longer. Theodore Davis and Emma Andrews were frequently looking to hire these hopefuls to work with them. Others, like Petrie, liked to stay in a tomb at Giza when he was working there with Mena House just a few steps away (not that he would have stayed there). Early on, Mariette had built a dig house at Saqqara and Carter stayed there when he was the inspector.[67] Carter also stayed at Mena House and other places in Cairo when he would take a break from the sites. For those, like James Breasted, who focused their time in Cairo on Coffin Texts and copying other scripts at the museum, along with planning the Epigraphic Survey and reconnaissance trips through Mesopotamia, centrally located city hotels, pensions, and private apartments were ideal. Each archaeologist who stayed, dined, and talked at the hotels in this section had one or more of these activities to attend to.

Upon coming to Cairo, more established archaeologists would stay in two main hotels where they would have close contact with their colleagues as well as with other travellers: Shepheard's and the Continental. Charles Wilbour, Theodore Davis, and Emma Andrews all stayed in these high-end hotels; James Breasted and Howard Carter were able to stay there once they had wealthy patrons. Before they were professionally established, or if they did not have a big budget, archaeologists would stay in smaller, cheaper hotels, such as the Hotel du Nil and Eden Palace. They used the larger, more well-established, higher clientele hotels as meeting places, but also as staging areas for getting jobs and establishing their professional networks that would propel them to positions that would bring them the budgets to stay in nicer hotels. Even Petrie was known to visit Shepheard's because he knew that 'people turn up in a marvellous way here'.[68]

This chapter has, thus far, explained Cairo and the general activities that visitors – both tourists and archaeologists – could expect to be doing while staying in the city. I will now focus on specific hotels in Cairo and the archaeologists that used these particular spaces to build their networks, find jobs, and establish the Western discipline of Egyptology exclusive of Egyptians. These hotels became important nodes in the cognitive topography of the city, excluding all but the wealthy, the connected, the hopeful, and the male Egyptologists. They also became truth spots, where interactions with the general public as well as professional connections were lent credibility before they moved out into the field site. As I investigate the careers of largely British and American Egyptologists, we will see that, as their status changed in the field, so could and did their hotel choices. Their hotel choice thus impacted who they were able to interact with, and that would have an impact on not only their personal careers, but also on the discipline of Egyptology through the early twentieth century.

Fledgling hotels and rookie scientific networks

Not a lot is known about the Guthrie Roger who arrived in Cairo in February 1893. He came to work on a project with Howard Carter for Percy Newberry and the London-based Egypt Exploration Fund (EEF) in the Delta. Carter had trained with Petrie in 1892 at Amarna, but was not 'built of such austere stuff' as Petrie, so he decided to shift his work to a new site and with a new lead archaeologist.[69] In the fall of 1892, he started working with Newberry; in early 1893, the EEF pulled Carter from his work with

Newberry in El-Bersheh, and moved him to Simbelawîn in the Delta.[70] Roger came to work with Carter on the new site, which, it was hoped, would produce well-preserved papyri. In the end, they were unable to excavate the site due to weather and the inability to get a permit, and their work ended on 13 April.[71] But before all this work began, Roger first arrived in Cairo to find that 'there was no room at the inn (the Royal Hotel) to which he had been instructed to proceed'.[72] Roger wrote to Newberry: 'I arrived here last night and, as I did not find the bus of the hotel Royale at the station, I have taken up my quarters at the Hôtel de l'Orient. I am informed that the Hôtel Royale is full.'[73] As Carter was not yet in the city, Roger told Newberry that while he waited for Carter he would do what all people did in Cairo: 'I will make the most of my time here'.[74] He toured around the city, saw some sights, and did a little shopping while waiting for further instructions. It is clear that he was a budding Egyptologist, working with the EEF, and he had likely hoped that this first season on the job would lead to a career, as it had done for Carter. However, he grew tired waiting for word from the Department of Antiquities. He wrote to Emily Paterson, General Secretary of the EEF:

> Thanks to a period of inaction, bodily and even mental, such as I have never had in my life before, I am nearer positive ill-health than I ever have been up till now. I do not complain on account of Oriental procrastination which I was prepared for, but I think the present situation might have been avoided. I shall not stay here longer than a few days more; if nothing definite happens in that time I will return home. The present position is too insufferable to be prolonged.[75]

No direction came and he was forced to return home to Scotland. Other than that, not a lot more is known about him or the rest of his career. Most of the discussion about him is in the context of his work with Carter in 1893. Roger, as others, relied on Cairo and the Hotel d'Orient to be a place of opportunity, but he ran into bad administration and never returned to Egypt.

The Hotel d'Orient was built in the 1830s or 1840s and sat on the northern side of the Azbakeya Gardens, not far from the train station. It was opened by one of Muhammad Ali's former cooks and offered sixty rooms to guests.[76] Sir John Gardiner Wilkinson's *Hand-book for Travellers in Egypt* from 1847 mentioned the hotel as an option outside of Hill's British Hotel (later Shepheard's), but there were so few hotel options at that point that Wilkinson talked far more about private homes to rent than hotels.[77] In the 1885 edition of Baedeker's *Egypt*, the Hotel d'Orient was described as having 'good cooking and moderate charges'.[78] It also had bathrooms and reading rooms. Being on the centrally located Azbakeya meant it was a good location for getting to and from the shops in the Muski and the Egyptian Museum, which at the time was just down the road in the Bulaq area. By 1895, the Hotel d'Orient had disappeared from the list of hotels in guidebooks, so it is likely that it closed. It had been a favourite of Petrie's, when he was forced to stay in a hotel. It was cheap, close to the train so he did not waste time, and close enough to the supplies he needed.[79]

Just a few years after the Hotel d'Orient closed, the small, short-lived Eden Palace was built in 1899. Due to its use by the British military during the First World War, it closed as a hotel in 1920. The building is still in Cairo, and the name plate still exists on the top of the building.[80] The Eden Palace Hotel was built on the northwest edge of the Azbakeya, a block away from Shepheard's and the Continental, which were on the northwest and southwest corners, respectively. Baedeker's touted it as having '145 rooms, lift, steam-heating, [and] frequented by British and American travellers'.[81] Visitors choosing the Eden Palace with rooms facing the Gardens would have a view of the trees, flowers, and grass when looking out the window but would pay almost half the price as those staying in the more expensive hotels.[82] The price cut may have been a matter of geography, though. Humphreys notes that while the front entrance of the hotel faced the lush Gardens, going out the back door would take travellers to Wagh al-Birket, an area which was known for its proximity to sex workers. He argues that this location likely did not help the Eden Palace's reputation.[83] What the location, and its subsequent reputation, did do for travellers was make staying in the hotel affordable. Those who would stay at the Eden Palace were a mix of tourists, business people, and those seeking the desert air for health, while on a budget, the so-called lung trippers.[84]

It was well known at the time that people suffering from tuberculosis, also known as consumption, or other respiratory diseases would find relief from their symptoms in Egypt.[85] Many wealthy, and not so wealthy, people would escape the damp, coal smog, winter air in Europe in favour of a change to the fresh, dry desert air.[86] The drier air, ideally, would stop the coughing and slow the progress of a then-incurable disease. Many who came to Egypt looking for reprieve for their lung ailments were quite wealthy. Some would stay out at the Mena House on the Giza Plateau, like Carter chose to do. Others would go to Helwan, the health resort town just south of Cairo.[87] Others, with much more money and time, might take a trip up the Nile and back. The Eighth Duke of Devonshire, Spencer Compton Cavendish, did just that from 1907 to 1908 and brought along his family and his physician, Ferdinand Platt.[88] They were not particularly interested in becoming professional Egyptologists or making a return visit, but they brought back antiquities and took advantage of their status to see all the sights along the way. In January of 1908, while moored for the second time in Luxor, the Devonshire party met Howard Carter and were able to get tours of some of the tombs he was working on. Earlier, they had met Winston Churchill, then the undersecretary of state for the colonies, who was inspecting Britain's East African colonies. They took the essential wealthy traveller version of the Nile journey, which I will detail in the next chapter, but, like all lung trippers in this period, they could not beat the tuberculosis that was slowly tearing apart the Duke's lungs. The Duke died on 24 March 1908 in Cannes, France, at the Riviera Hotel, on the way back to his Derbyshire home from Egypt.[89]

Although a number of those who travelled to Egypt for their health were indeed quite well off, many of them were not. When they arrived, they needed to find a job or their time in Egypt, as well as their health, would be even more at risk. Another one of these lung trippers was the budding Egyptologist, E. Harold Jones, who was hoping to start a new career as an artist and copyist.[90] Jones was a young Welshman who trained at the Royal College of Art. By 1904 he was suffering the effects of tuberculosis, and, hoping to begin a new career as well as recover his health, he joined an expedition to Egypt with Liverpool Egyptologist John Garstang.[91] The ever-popular 'change of air' seemed to work for Jones almost immediately. On his arrival in Egypt in 1904 as part of Garstang's 1903–04 season at Beni Hasan, Jones wrote home on letterhead from the Continental, that on his first morning in Egypt 'I got up at 9:30 & was relieved to find my morning cough had gone.'[92] Even though his first letters were from the Grand Continental, he probably did not stay there. He ate lunch there and met people at the Continental, but most of his letters from Cairo come from the Eden Palace, which was the cheapest option, he said, because it was full of lung trippers.[93]

For the following season of 1904–05, Garstang hired Jones again as the artist for the Liverpool expedition. Throughout the early twentieth century, many expeditions hired artists on site due to the difficulties of photographing sites accurately.[94] They began work at the sites of Hierakonpolis and Esna, but a disappointed Jones had been deserted by Garstang to do most of the work while Garstang was in other locations.[95] Jones worked for Garstang for a few years, but grew tired of the isolation (Beni Hasan, for example, was twenty kilometres south of the next largest town, Minya), and the fact that Garstang was often off site, either taking care of business on other Liverpool sites or, as Jones claimed a few times, playing too much golf.[96] Jones both longed for and dreaded going back to the damp air of England each summer to see his family, and was desperate to stay in Egypt. Throughout what would be his final season with Garstang in 1905–06, he took advantage of the wide network of Egyptologists staying in hotels all over Luxor and Cairo in order to find a new job. In May of that final season, he wrote home to his parents from Cairo:

> I went to the Continental Hotel front to tea – all alone thinking I'd possibly see somebody I knew passing. I had not long to wait for [a friend] was driving past & came up to speak for a few minutes & then – how glad I was to see him – I saw dear old [Percy] Newberry ... walk up – he hadn't seen me – but I called him. I didn't know he was in Cairo still & we spent the rest of the evening till dinner together.[97]

Jones' tactic of waiting for somebody he knew to pass by worked as he expected it to. Newberry, also based at Liverpool at the time, was actively digging in Egypt at this point so he was able to get Jones some work that would keep him in Cairo a little longer. Jones did not know Newberry was still in Cairo, because usually, by May, Western Egyptologists were back at their home institutions to avoid the heat of Egyptian summer and to work on writing up and exhibiting their work. Jones concluded his letter to his parents, 'I may stay a night with the [George] Reisners & Howard Carter the artist has asked me to his place just outside Cairo. I am so lucky in making friends – I have heaps & heaps of work to do here & could spend months in the museum.'[98] And he would do just that.

Just a few days before he met Newberry at the Continental, Jones had written to his parents from the Eden Palace Hotel, which is probably where he was staying. He was still looking for work at that point, and Egyptologist and artist Norman de Garis Davies had offered him £300 per year 'to do nothing but copying [sic] tombs ... But I refused – I couldn't stand the work at his pace all the year round & in fact with more as the other Davis [that is, Theodore] pays me better his work taking up but about six weeks or two months in the year.'[99] The Americans Theodore Davis and Emma Andrews would successfully woo Jones away from the gruff, low-paying Garstang. But even the excitement of working at Amarna with de Garis Davies could not draw Jones away from the money and glamour of working with the Americans. Davis had provisionally employed him in March of 1906, through a meeting in Luxor, to draw plates for Davis' publications of his excavations in the Valley of the Kings.[100] By the opening of the following season, November of 1906, he was working full time for Davis and Andrews. And in February of 1907 they found a large shrine, covered in gold leaf, that he copied. Jones was talented, of that there is little doubt. Because of his talent, he was able to successfully use hotels as meeting spaces for impromptu interviews in order to boost his career. He had got himself more than a few jobs using this method, and we will see his career take off in Luxor (Chapter 4, below).

Building central hotels and network nodes

James Breasted did not come from a wealthy family. He was born on the prairies of Illinois and by the age of twenty-two he had found his niche studying ancient languages. He was convinced by his Hebrew professor in Chicago that he should study Egyptology, which was then a 'vacant field' in the United States.[101] He first went to Yale then, in 1891, to the University of Berlin to study with one of the foremost scholars of the day, Adolf Erman. He earned his PhD in Egyptology in 1894 – the first American to receive this distinction. Although he had completed all the university requirements to earn the title 'Herr Doktor der Philosophie Breasted, Hochwohlgeborner', and had a job waiting for him at the new University of Chicago, he still had to make the journey to and through Egypt to establish himself as a true professional.[102] As with many field sciences, in Egyptology a degree alone did not give Breasted professional standing. Erman thus urged Breasted to go to Egypt 'for the sake of his health and scientific future', and gave him an important task: collating inscriptions in the Egyptian Museum in Cairo for a massive dictionary Erman had begun work on.[103]

Despite the important job he had waiting for him in Chicago and the research he would do for Erman, he was newly married to Frances, an American woman he had met in Berlin, and the couple had little money between them. Breasted convinced William Rainey Harper, his old Yale professor and new president of the University of Chicago, to grant him one term's leave of absence at full pay so that he could establish himself as a professional Egyptologist. His parents gave him some extra money to go on this trip, knowing it would form the foundation for the rest of his career. Even with the pay and extra family funds, the Breasteds arrived in Cairo in November of 1894, hearts full of excitement but purses near empty. They went to one of the least expensive hotels they could find, which Petrie may have suggested to them.

Frances later recalled:

> With our slim though united purses … it never occurred to us to go to the already famous Shepheard's Hotel, or to the Continental Hotel. We went instead to the old Hotel du Nil in the middle of the Mousky [sic]. This seemed to us an oriental paradise. The balcony of our room looked down upon a beautiful garden surrounded by a high white wall and filled with tropical vegetation and tremendous palms, in the midst of which played a fountain. It was a cool, enchantingly peaceful oasis in the dust and turmoil of the native quarter. From an observation tower atop of the hotel, to which we mounted at sunset to look out upon the city so new and full of romance to our western eyes, we counted more than two hundred minarets. But even the Hotel du Nil was too great a tax upon our slender resources.[104]

Breasted did not record if there were any other archaeologists staying there at the time, but it is possible they crossed paths with soon-to-be archaeologist Margaret Benson and her brother Fred who also stayed at the Hotel du Nil in late 1894. Benson was another wealthy lung tripper who sought Egypt's dry climate to ease not only her lung maladies, but her emotional ones, too. She did not have tuberculosis, but her lungs were weak and she suffered from depression. The Benson siblings only stayed there for a few days, while on a Thomas Cook's trip, and we will learn more about their travel and archaeological activities in Luxor in Chapter 4.

Established in 1836, in the middle of the native quarter of Cairo, the Hotel du Nil was a converted Arab house with a large arched gateway that opened up into a courtyard filled with palm, banana, and orange trees (Figure 2.4).[105] In 1846, Harriet Martineau recorded that both the du Nil and Shepheard's were full when she arrived in Cairo; in fact, before 1882 – that is, before the main period covered by this book – the du Nil was one of the main hotels in Cairo, and a number of archaeologists and tourists stayed here. Even though the hotel was relatively new in 1858, Briton William Howard Russell remarked of his stay at the du Nil on his way to India, that it was located '[i]n the dark, among the dogs, through lanes and alleys of infinite closeness, nastiness, and irregularity [...] It was called the Hotel du Nil, and it well deserved the name, for we could get nothing to eat, not even a piece of bread, when we arrived.'[106] Twenty years later, in the 1878 Baedeker's edition, they described the du Nil as being 'in a narrow street of the Muski, the main artery of traffic; a good house, though uninviting externally, with a pleasant garden'.[107] It also had baths, and only cost about sixteen to twenty francs per day.[108]

That it was a favourite of Egyptologists early in their career is without question. Some liked to stay there because they preferred not to be bothered by tourists who usually stayed in the more European areas. Most of them liked it, as the Breasteds did, because they could not afford to stay anywhere else without patrons, grants, or larger salaries. They would have been close to the antiquities dealers, and they enjoyed being away from the tourism hustle and bustle. In the 1880s, as the Breasteds did, Flinders Petrie found the du Nil too expensive, and preferred to stay in the Hotel d'Orient near the train station for a day or two before moving out to the tombs on the Giza Plateau, to be closer to his work. By 1892, his new position at University College London (UCL) meant that he could afford to stay in a nicer hotel. When he brought his new wife Hilda to Egypt on their honeymoon in 1897, they stayed at the du Nil and it became their hotel of choice in the city. This is not to say that the Petries would have talked to many colleagues if they saw them. The Petries were hardly good company, even in their early days, choosing work over festivities most of the time. Petrie therefore used his lodgings as bases for his museum work, getting permissions for excavations, and getting supplies.[109]

Wallis Budge, Keeper of Egyptian and Assyrian Antiquities at the British Museum from 1883 to 1924, travelled to Egypt a number of times in his

Entrée de l'Hôtel du Nil.

Figure 2.4 Entrance arch to the Hotel du Nil, from hotel brochure c. 1895.

official duties. When he arrived at the Hotel du Nil in 1886, he recalled that he

> found assembled Walter Myers, Henry Wallis, Greville Chester, a couple of dealers, and several other men who were interested in Egyptian antiquities ... In the course of a long evening's talk I learned many things about the 'antiquarian politics' of Cairo, and I found the information I received from the company generally most useful in later days.[110]

There was plenty of this table talk to go around over the years with many other archaeologists who stayed here. Budge's tasks in Egypt tended to be focused on acquiring artefacts, which meant that, along with excavating and working with archaeologists, he bought a lot of antiquities and sent them to the British Museum in London. Budge worked to establish relationships with dealers to ensure the continued flow of artefacts into the British Museum in the late nineteenth century.[111] He would not have got these lessons from any sort of institutional archaeological training. In fact, the only way he could learn the politics of buying from dealers all over Egypt was from listening to his colleagues who had had to learn these practices by doing them. Buying from dealers was a legal (if questionable) practice in those days, with local dealers having to compete with the museum shop itself.[112] Although the practice has largely been curtailed over the last century, Egyptologists, curators, tourists, and private collectors alike would purchase artefacts, many of them authentic, to build their collections back home.

By 1895 the Hotel du Nil had been rebuilt, largely due to general disrepair.[113] An updated description comes from Auguste Wild, a Swiss hotelier who managed the Savoy Hotel and recalled the du Nil in his 1954 memoir, *Mixed Grill in Cairo*.[114] He arrived in Cairo in 1898, and described the hotel as he saw it then:

> Another hotel of old standing was the Hotel du Nil, which was situated at the heart of the native quarter of the city. It was approached through a dark, inconspicuous archway which led, surprisingly, to a picturesque building in the midst of an old and luxuriant garden. It boasted no modern amenities in the way of sanitation or other conveniences; nevertheless, it continued for a long time to be the resort of Egyptologists and other savants and scholars, who preferred its oriental seclusion to the gaieties of more fashionable hotels.[115]

Throughout the years Baedeker's *Egypt* continued to recommend the small hotel for tourists, but not as highly as Shepheard's or the Continental. The 1895 edition of Baedeker's guide told visitors that the Hotel du Nil was 'in a narrow street off the Muski, recently rebuilt with a pleasant garden, terrace, and belvedere, good cuisine'.[116] By 1902, a night cost fifty piastres during the normal excavation seasons (October to May) and forty-five piastres otherwise, with electric light costing one piastre extra.[117]

In 1905, the Breasteds returned to Egypt and to the Hotel du Nil, this time with their eight-year-old son Charles. Their purses were equally as slender as they had been ten years before, but Breasted, still in the process of building his scientific network in the field, used the hotel as a professional base.[118] Charles recalled in his biography of his father, *Pioneer to the Past*, that from their home-base hotel in Cairo they 'made the rounds' to see the sights, the sounds and smells of the old bazaars ('smelling of incense, musk, and sewage'), antiquities dealers, British consuls, and the museum.[119] The museum visit was to see Gaston Maspero, then the Director of the Egyptian Antiquities Department, in order to secure a permit to copy the ancient monuments of Upper Egypt.[120] Charles later described the setting as 'crowded with open boxes of recently excavated antiquities … Here, like treasure in a cave, was everything imaginable, all the things my father had told me I myself could perhaps dig up with a small shovel!'[121] Charles later on remembered little of the meeting except that his father and Maspero 'talked their shop'.[122] Maspero did not grant permissions that year for all the ancient monuments, but instead told Breasted to 'select a special district in Nubia and to work it out before asking for a second one'.[123] He also told Breasted not to clean or excavate any of the sites; Breasted agreed and just like that, the permissions were granted. Frances was tasked with buying up the last-minute items for the trip up the Nile in order to carry out the work.

The final time the Hotel du Nil was on the Baedeker's map was in 1902; it was still a hotel in 1905 for the Breasteds to stay; but by 1906, the du Nil was gone. In her diary, on 28 November 1906, Emma Andrews noted that the hotel had 'been torn down, and is rebuilding', but there is no evidence that it was built on the same spot.[124] Its name was swallowed into the Hotel Bristol and du Nil, probably by an owner hopeful to receive the du Nil's clientele.[125] The 1908 Baedeker's map still had the building highlighted, but no name marked the spot. The name of this establishment disappeared in archaeologists' writings and discussion, as well as from the guidebooks.

The fact that the du Nil had closed by the end of 1906 did not matter much to Breasted, who by then had established himself as a leading Egyptologist. He was also one of the main beneficiaries of the John D. Rockefeller Foundation, and Rockefeller himself personally, who would be his biggest financial and political supporters throughout his career. While the Hotel du Nil was a small, older hotel, it formed the foundation of the work that archaeologists such as Budge, Petrie, and Breasted relied on early in their careers. There were, mostly, Egyptologists and scholars who stayed in the du Nil, which would have had an effect on what could or would be discussed or been allowed at the dinner table. However, without a central space like a main dining room or terrace in the hotel where people could regularly meet, the Hotel du Nil's importance was as a place to stay while working and for new visitors and archaeologists to get to know the shape of the city and the layout of their networks, rather than as a working space where already-present networks were strengthened. The Hotel was a crucial place for many Egyptologists to start their work, meet other important travellers so they could build their networks, and make themselves into the scientists they needed to be. In their early days in Egypt, they were establishing their bona fides as Egyptologists and the Hotel du Nil was cheap enough and had enough like-minded visitors to help them do this.

Colonial strongholds: Egyptologists' work in Western hotels

The two most famous European hotels in Cairo were also central to the development of Egyptology in the late nineteenth and early twentieth century. With myriad less-famous hotels and the network-building processes they fostered in forming the foundation blocks of our grand Cairene intellectual landscape, we have arrived at the two shining, golden capstones: the Grand Continental Hotel and Shepheard's Hotel. Each of these hotels have merited several book-length treatments, and they could each be their own chapters, but I have included them here in order to keep them within the appropriate context of Cairo.[126] Shepheard's and the Grand Continental, also previously named the Continental-Savoy or the Continental, were the hotels of choice for many people once they got to Egypt. The hotel spaces themselves were colonial strongholds which helped to shape the discipline of Egyptology for over a century, and the people who inhabited and worked within their walls benefitted from these spaces to bolster and enhance their power in the country and in the discipline.[127]

Wealthy patrons stayed in these hotels for comfort and luxury. While many Egyptologists could not afford to spend many nights at the more expensive hotels, they still met people there. They were also central places to meet their friends and their archaeologists; archaeologists themselves would often stay there only once their careers had been established, either through work with a personal patron or through institutional prestige and support. Wealthy Americans Charles Wilbour, Emma Andrews, and Theodore Davis often stayed at Shepheard's; the wealthy patron of Howard Carter, Lord Carnarvon, usually chose the Continental (Figure 2.5). James Breasted alternated depending on his needs and his mood. There are many others whose stories I will tell throughout this chapter.

Shepheard's Hotel: colonising a country and a discipline

The first hotel officially to be called Shepheard's was opened in July 1851, but Samuel Shepheard had been managing hotels in Cairo since 1842.[128]

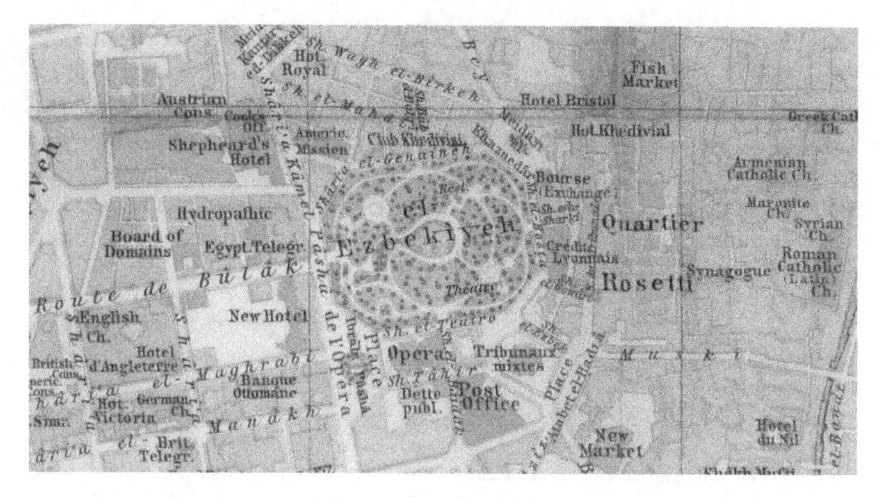

Figure 2.5 Baedeker's map, detail of hotels in Cairo, 1895.

First he managed the British Hotel under its owner, a Mr Hill. Known sometimes simply as Hill's, the British Hotel became the first building called Shepheard's by its patrons – because of the manager's popularity among visitors – and it faced the Azbakeya Gardens. The earliest iteration of this famous hotel seems unsophisticated in comparison to its more opulent offspring over forty years later. In the 1850s, the 'balcony in front was narrow, uncovered and paved with stone slabs', its rooms were uncomfortable and small.[129] The view from the front stretched 'across to the Hotel d'Orient for many acres of space was to be seen the old Ezbekiyah, planned and planted by Mehmet Ali the primitive Hyde Park of Cairo'.[130] The hotel underwent a number of changes, first being sold by Shepheard in 1860 and then being revamped after a fire in 1869, just in time to host visitors who arrived to see the opening of the Suez Canal.[131] In his *The Innocents Abroad* from that same year, Mark Twain complained about the state of the hotel, including the old, worn carpets and bad lighting. But he had stayed there in 1867, before the renovation. In 1874, Amelia Edwards famously remarked in the first line of her *A Thousand Miles Up the Nile*:

> It is the traveller's lot to dine at many table d'hôtes in the course of many wanderings; but it seldom befalls him to make one of a more miscellaneous gathering than that which overfills the great dining-room at Shepherd's [sic] Hotel in Cairo during the beginning and height of the regular Egyptian season.[132]

The hotel became so famous among visitors and their audiences that it continues to live in the memories and imaginations of Egyptologists, historians, and novelists.[133] Anthony Trollope recognised as early as 1867 in his 'An unprotected female at the pyramids' that 'the English tongue in Egypt finds its centre at Shepheard's Hotel'.[134] Shepheard's appeared in movies such as the 1934 film *The Camels are Coming*, where viewers can catch agonisingly short glimpses of the famous terrace where people drank tea, dined, and talked.[135] Even Elizabeth Peters' grand, if fictional, Egyptological couple, the Peabody-Emersons, stayed at Shepheard's. In the first book of the Amelia Peabody series, *The Crocodile on the Sandbank*, Peabody finds herself and her travelling companion, Evelyn, at Shepheard's in 1884. She recorded this in her fictional journal, echoing Edwards' memoir:

> Everyone stays at Shepheard's. Among the travellers who meet daily in its magnificent dining room one may eventually, it is said, encounter all one's acquaintances; and from the terrace before the hotel the indolent tourist may watch a panorama of eastern life pass before his eyes as he sips his lemonade.[136]

Many real travellers enjoyed these benefits over the years, but even with the post-fire renovation in 1869, Shepheard's was an old, dirty, 'murky' hotel by the time it was torn down in 1890 and completely rebuilt, opening in 1891 to international fanfare.[137] It is this version of the grand hotel I will be talking about here, the building and space that many knew as Shepheard's in the late nineteenth century, which was not the hotel that its founder would have known.

It has long been argued that the first Egyptologist associated with the new Shepheard's was G. Somers Clarke, also an architect, who designed the new hotel.[138] Clarke had worked with James Quibell at the site of Hierakonpolis and had 'superintended the restoration of several temples and other buildings in Egypt'.[139] However, thanks to Tarek Ibrahim's serendipitous discovery of a rare primary source about Shepheard's on eBay in April 2014, and the subsequent publication of the archival information that came of the source, the architect is now recognised as German architect Johann Adam Rennebaum.[140] Rennebaum designed the hotel as an open rectangle surrounding a courtyard, open to the sky, containing fountains, tropical trees, and birds; there was also a southern annex.[141] Important for our visitors was that the famous terrace replaced the old balcony, and there were a number of restaurants and bars inside to meet and eat.[142] This renovation, which only took five months to complete, made Shepheard's the first hotel in Africa to have electric light. There was also a steam laundry in the basement. The newly renovated hotel contained 340 bedrooms and 240 bathrooms, which was relatively luxurious, considering that about two bathrooms for an entire floor of hotel rooms at the time was seen as generous.[143] By 1898, the great Entrance Hall was enlarged to include the soon-to-be famous Moorish Hall, an octagonal-shaped room with a dome of coloured glass as the ceiling.[144] Historian Nina Nelson noted that when 'the building was completed it ranked as one of the most elaborate hotels of the day'.[145]

Shepheard's was not only an important space for Egyptologists to meet, making the hotel a central truth spot on the ephemeral intellectual landscape of Cairo, but it was convenient for them to stay there because they regularly used the shopfronts that were in and near the hotel. There was a Thomas Cook & Son office in an annex of the hotel, as can be seen on Rennenbaum's original plan.[146] The Egypt Exploration Fund (EEF), based in London, regularly sent money to their Egyptologists all over Egypt from London through Cook's offices.[147] John Cook, the son of Thomas Cook and president of the company from 1892, was a member of the EEF by at least 1889. Cook's Cairo office also displayed and sold EEF publications to their tourists.[148] The EEF also contracted with Cook for travel, and they often received tickets for travel at a deep discount, up to half price.[149] Most mail for these Egyptologists was sent to them in care of Cook's offices in Cairo, Luxor, or Aswan and would be held there until someone was able to pick it up. The frontage of the hotel featured space for shops, including a post office, an antiquities dealer, and Diemer's bookshop. In November of 1904, Diemer's shop expanded and the owner, Felix Marschner, built special shelves specifically for the books and site reports published by the archaeologists working for the EEF.[150] The connection between Egyptology, Thomas Cook's, and Shepheard's Hotel was clear.

Just outside the front doors of the hotel, and framing the front of the building, sat the famous Shepheard's marble terrace, where so many Western residents of this hotel and Cairo in general would come to sit and talk (Figure 2.6). One visitor from St Louis, Missouri, echoed in 1898 the prevailing sentiment of visitors for the next sixty years: 'sitting upon the marble terrace, [we] watched the moving panorama in the narrow street; where so much of Oriental life can be seen without any effort on your part, for nearly everybody and everything comes to or passes Shepheard's in the course of the day'.[151] This view of life included donkey boys offering rides to the pyramids, flower boys, serpent charmers, wedding parties, funeral processions, pilgrimages, and more.[152] It was into this opulent setting that everyone walked when they entered Shepheard's in its heyday, from 1891 to the early 1950s. Shepheard's was open all year, but the official Cairo tourist season began when the dances at Shepheard's started, sometime in late September or early October. The terrace would fill up quickly and rooms were difficult, if not impossible, to get if you had not booked early. However, anyone could come to meet on the terrace, and pretty much everyone did.

The wealthy American travellers Emma Andrews and Theodore Davis chose Shepheard's as their regular lodgings each year while they waited for their dahabeah to be readied. Emma Andrews was described as 'intelligent, cultured, and unconventional'.[153] By the time she met millionaire lawyer, robber baron, and antiquities collector Theodore Davis, she was already a

Figure 2.6 Egyptian Hotels Ltd, Cairo. Shepheard's Hotel. Exterior, c. 1920.

wealthy widow.[154] Davis, wealthy lawyer and colleague of Charles Wilbour (see Chapter 3), was a major figure in Egyptology from the 1890s until his death in 1915. Some say Andrews was Davis' mistress, while others argue she only became his close friend and travelling companion when his wife Annie showed no interest in Egypt. Some historians argue that the nature of their association does not matter historically, and in some ways they are correct. However, the nature of their relationship is important in terms of seeing them as an influential scientific couple. Their partnership was key to Andrews' ability to go to Egypt, to work with Davis, and to be seen as his equal. In truth, it is uncertain what kind of relationship they had, but it bears investigation. For almost twenty-five years – between 1889 and 1913 – Andrews was not only Davis' travelling partner, but an eager collector, diarist, and knowledgeable archaeologist in her own right. From their first trip to Egypt to their last, she kept a detailed diary about their time on site, and on their dahabeah, eventually commissioning the *Bedauin*.[155]

As many wealthy vacationers did in Egypt at the time, Andrews recorded her thoughts about Egypt, its people, the sights up and down the river, the weather, the river itself, and most of all – the antiquities. Each year, the couple brought family or friends with them, and, while on the Nile, they preferred to entertain on their dahabeah rather than in cafés or restaurants. Their activities on their first trip up the Nile, beginning in December 1889, differed little from what archaeologists did during their time in Egypt. They arrived in Alexandria early on a December morning, stayed only to eat lunch at the Hotel Abbat, and then caught the train to Cairo, where they arrived at around 9:00 pm.[156] They went directly to Shepheard's Hotel. As Andrews described it, on their arrival at Shepheard's, they were 'installed in delightful rooms into which the sun looks all day – and out of our windows we look upon a spacious court, with palms and tropical trees, among which the great black and grey crows are flying and cawing all the time'.[157] The pair had been in Egypt before, but this was the first time they would make a trip up the Nile.[158] Although they would not begin their Egyptological activities for a few more years, they would continue to choose Shepheard's whenever they were in Cairo. They knew it was already a peak in the cognitive topography of Cairo, so they used the hotel's fame, and their own reputation as wealthy donors, to woo and hire archaeologists, artists, and other crew to join them on their excavations in the King's Valley in Thebes. Harold Jones, while he did not stay there, dined at Shepheard's with the pair and that was how he got his first job with them.

More well-known Egyptologists later in life were often too poor to consider staying at Shepheard's when they first arrived in Egypt. James Breasted, as we have seen, had some institutional support on his first two trips in 1895 and then 1905 but no real money for a good hotel; he stayed at the Hotel du Nil both times. As much as he disliked Cairo, with its bustling streets and lively social scene, he understood that the city was a centre of power and he needed to meet with people in power to get their support.[159] After the First World War, Breasted was one of the first Western archaeologists allowed to go back to Cairo to begin a new project, thanks to funding from John D. Rockefeller and support from the British military.[160] After a number of delays in London because both the Americans and the British feared that the area was too politically unstable, he arrived in Egypt in late October, using Cook's agents to get through customs in Alexandria then and into Cairo. He was flush with $20,000 per year for five years from Rockefeller and emboldened by his excitement for his new Epigraphic Survey project, so, uncharacteristically, Breasted began his stay in Cairo first at the luxurious

Continental.[161] He found the place too loud and not private enough (he did not have his own bathroom), so he quickly moved to Shepheard's 'where I have a room opening on the garden, with a bath connected'.[162] A few days after he moved there, he found Shepheard's was so popular and he met so many of his friends there that he complained to his family, 'you would think that I might be fairly able to get a few minutes by myself here in Cairo, but it is curious how I am unable to do so'.[163]

His work in the museum was productive, though. He continued working apace for the Coffin Texts project, copying the museum's unpublished inscriptions, and his own Epigraphic Survey.[164] He preferred to work in the museum undisturbed, and raved to his family when he could finally 'settle down at the museum, with a table, a chair and a step ladder' to copy inscriptions.[165] He was also purchasing antiquities for the University of Chicago's new Haskell Museum, and a number of other museums in the United States which had given him money to buy objects on their behalf.

In early November 1919, on his way back to Shepheard's from the museum, he wrote to his family that 'the good Golénischeff overtook me'.[166] This was Vladimir S. Golénischeff, a Russian Egyptologist whom Breasted had helped gain a position at the museum, cataloguing the hieratic papyri there.[167] They walked and talked, in their common language, French, but Breasted wrote that 'as we issued into the old Sharia Kasr en-Nil, I caught a glimpse of the marvelous afternoon light on the slopes of grand old Mokattam looking down on the minarets of Cairo; – I sighed, my mind wandered and for a moment or two I was entirely unaware of what he was saying'.[168] They ended up back at Shepheard's, where, before the Russian Revolution, Golénischeff 'had once occupied the finest suite in the house winter after winter'.[169] Ultimately, the two linguists sat in the Shepheard's writing room to discuss grammatical issues in the Egyptian scripts, but Breasted did not share the details with his family, so we do not know exactly what they said. However, it is probable that their conversation impacted Breasted's work on the Coffin Texts as well as Golénischeff's work on the papyri he was engrossed in. As Golénischeff left, an American Colonel Allen, the military attaché of the American Agency who lived at Shepheard's, came up to Breasted and asked him to determine the value of the boxes of scarabs he had come to own. Breasted told his family: 'I had finally left him with the assurance that he had a fine assortment of scarabs worth about 75 cents a bushel as produced at Gurneh (across from Luxor).'[170]

Despite the fact that he could easily meet with and talk to friends and new acquaintances at the hotel, both Shepheard's and Cairo were causing some social and professional productivity problems for Breasted. He lamented to his family that he struggled with the mix of 'social and official strings which have to be kept pulled all the time'.[171] If he ran into people unexpectedly, he usually took advantage of the time to talk shop, but it did not mean he enjoyed having to do that. He continued, explaining his current situation:

> for example, Guy Brunton and his wife are out on the terrace there now, having dined here this evening with friends, and I must go out and talk with Mrs. Brunton who paints wonderfully and is doing some portraits of the Pharaohs and their queens which I want to use for my new edition of the history.[172]

It is not clear if she did any of the illustrations for Breasted, but just five years later, Winifred Brunton's paintings were indeed published in a book entitled *Kings and Queens of Ancient Egypt*, which was possibly conceived of in that conversation and which contained a foreword written by Breasted.[173]

To further complicate his time in Cairo, Breasted had been given a dinner by a Mr Gary, with two unnamed but wealthy New Yorkers (not Andrews and Davis), and Colonel Allen. There was a ball after that dinner, one of the famous Saturday balls at Shepheard's, which in 1919 would have been full of wealthy tourists, members of the military, and diplomats. Here Breasted met General Louis Bols, General Edmund Allenby's chief of staff, who was working out of the old Savoy Hotel. He regaled his family with stories of the evening:

> We had a very interesting conversation at once. It was like a scene from one of Mrs. Humphrey Ward's novels, this brilliant ball room, filled with the big men of the British Empire who are out on the frontiers doing things, and taking their relaxation in a roomful of beautifully dressed and pretty women, and doing it with great gusto and enjoyment; while all around the air is keen with rumors of impending trouble. It was indeed a fascinating experience to stand in a corner with one of the leading men in the situation, and watch the whirl of *American* dances, which after all we scarcely saw, as we talked of the big game of modern empire in the Near East and the grave dangers which French insistence on coming in and taking Syria, has introduced into the situation. It means the continuance of the world-war, and the French army, or its best men, are as much aware of the fact, as their Foreign Office is ignorant of it. At this juncture, the General's lady sent his A.D.C [aide de camp] to come and go home with her, and after wandering about among the gay uniforms for a few minutes to settle my mind, I left them all to dance away the night and turned in.[174]

Within a week of this meeting, all the social stress and noise started to bother Breasted, and he complained to his family:

> I have been very uncomfortable at Shepheard's: – noise, dust, bustle, society, monotonously tasteless food, of which one regularly overeats, etc., etc. My room was on the beautiful garden, to be sure, but it was on the ground floor, it was dark and somber if I kept the blinds down, and my things were not safe if I opened them. There were two open catch basins directly under my

> windows, from which I frequently caught the dank odor of ancient wash water, reeking with strong soap and filth.[175]

So, he moved out to a private Villa in Garden City, Cairo, which was a new quarter at that time. He said it reminded him of

> the old Egypt with its eternal mystery and charm, which are as strong in their hold upon me now as they were when we first looked out on Mokattam from the roof of the Hotel du Nil twenty-five years ago. [From my window] ... there is a grand prospect toward Mokattam, and I look from a balcony down the street to a broad stretch of the Nile from the library window.[176]

His move away from the central social and scientific space of Shepheard's did not have the simplifying effect that he hoped it would. In fact, Breasted himself had quickly become a central network node in Egyptology in Cairo. Further, being out in Garden City, he was much closer to the Official Residence of General Edmund Allenby. Allenby had just been appointed the High Commissioner of Great Britain to Egypt, the Sudan, and Palestine, which, Breasted said, simply meant that 'he is king of these countries'.[177] Breasted knew he needed to meet with Allenby, both because he had been given a letter of introduction to Allenby by A. J. Balfour, the British Foreign Minister, and because Allenby had been told by Balfour to give Breasted all of the travel support he needed.

Breasted introduced himself to General Allenby and soon entered into his confidence. Breasted wrote to his wife about their first meeting on 30 November that very evening:

> I had taken the first opportunity to say that I hoped to further the establishment of cordial cooperation between his country and mine in the future control of the Near East, and for that reason I would be glad of an opportunity to learn all the facts regarding the situation which it might be proper for me to know. He made no response whatever, but he at once began to talk.[178]

Allenby began talking about all of the political issues the British were having trying to manage the new Middle East. Breasted had found himself in the middle of international negotiations.[179]

A few days later, Breasted took the General and his wife on a desert excursion to the pyramid field at Abu Roash. Breasted reported to his wife that he met the Allenbys at the British Residence in Garden City for lunch 'at 12:45 (which is very early for English hours)'.[180] At lunch, Breasted and Allenby had talked about the recent Paris Peace talks, the absence of Iraq's Prince Feisal, and the disagreement between the British and French about what to do with the area. After lunch, they made 'a quick transfer to the Mena House [Hotel] by car, and there horses and a ride up the margin of the desert northward to the ancient causeway leading up to the plateau and the substructure of the pyramid of Dedefre, successor of Khufu (Cheops)'.[181] Breasted reported to his wife:

> The horses had not been ridden for a day or two and were full of ginger. They dashed off at a rapid canter along the desert north of Mena House and my camera was soon flying about most uncomfortably. Allenby noticed it and riding up to me insisted that I give it to him to hand over to one of his orderlies who was riding behind us [in a car] ... at length we had reached the top and stopped in the midst of a great stone yard where vast blocks of granite had been cut into shape for the pyramid. Allenby was very much interested and asked many questions.[182]

It is certain that they continued to discuss the matters of the peace talks and what Faisal might do, as Breasted told his wife that after the trip to Abu Roash, 'Allenby showed me the new map of Jerusalem, as it had been newly laid out to protect entirely the old walls and city from encroachment by any modern buildings whatsoever'.[183] This was the beginning of a long and difficult journey in which Breasted and his crew – Daniel Luckenbill, Ludlow Bull, William Edgerton, and William Shelton – became the first non-Arab people to cross the Mesopotamian desert since the founding of the new state.[184] The expedition had begun as a scientific expedition in 1919, but had immediately been swept up into the British Intelligence mission. Breasted's situation as an archaeologist-turned-spy was not unique: the British had been using archaeologists as spies since the 1870s.[185] However, the fact that he was an American gathering intelligence for the British was indeed extraordinary. He then took a three-month journey through the newly divided Mesopotamian and Persian region not only to map archaeological sites, but to be a spy for the British military.[186] As Breasted was about to leave for his journey across the desert, supported by Allenby and the British military, this was surely a topic of conversation for much of the time they were together. Egypt, like much of the Middle East after the First World War, was in some state of political upheaval.[187] The anti-British sentiment was simmering just below the surface and would often boil over into violent protests against the colonial power. But Breasted still had Egyptological work to do.

For all of the idyll and peace that (mostly) European travellers enjoyed over the years, Shepheard's was also a place for Egyptians to voice their dissent of the colonial situation. In 1910, students from the University of Cairo gathered around the hotel to protest former US President Theodore Roosevelt's recent address in Cairo. In it, he supported Britain's colonial policies within Egypt and argued that it would likely be 'generations before Egypt is able to govern itself'.[188] From the windows and terrace of Shepheard's, Western visitors watched the first anti-American protest in the Middle East take place outside on the street. Egyptians had long recognised what the West had wrought, and Shepheard's was a symbol of colonial power and a stronghold itself. To tear down Shepheard's would be a political act that would begin to burn down the power and authority that the West held in the region. Around forty years later, spurred by the same anti-Western, and specifically anti-British, sentiment, more Egyptians took to the streets on 26 January 1952. They burned or destroyed European-owned businesses, including burning down the whole of Shepheard's Hotel, long seen as a British outpost on Egyptian soil. A hotel with the same name was rebuilt in a different site, along the Nile, in 1957. It still stands in that spot, no longer holding the power or draw for Europeans that other hotels of the same name did.

As a symbol of Western colonial power in hotly contested Egypt, Shepheard's was one of the main places where these archaeological networks were built. Shepheard's was a central place to see and be seen for all of Cairo and Egyptological society; it was loud, crowded, and sometimes over-stimulating. Archaeologists and their patrons, like the Bruntons, Andrews, Davis, and Breasted, used Shepheard's for what they knew it was – a protected space in which they could plan their work and find people to help them carry it out. They used Shepheard's as a home base in which to construct their social and professional networks for the season and beyond. Artists and aspiring Egyptologists, like Jones, knew they could go to Shepheard's to find work and new groups in which they could work to be included and therefore continue to work season after season. Everyone could go there to meet with colleagues and friends to have tea and watch the world go by.

Carter and Carnarvon at the Grand Continental

One of the longest lasting hotels in Cairo was the Continental. The first building to occupy that specific lot of land was designed for British tourists as a stop-over on the Overland Route to India. Named the New Hotel, it opened in the same year that old Shepheard's had been renovated and the Suez Canal opened, in 1869. The New Hotel quickly became a competitor for the newly renovated Shepheard's as a place for tourists staying for an extended period of time. Just down the street from its more famous neighbour, it faced the Azbakeya Gardens, and had large verandas and arcades. It contained over 100 guest rooms and a billiard hall.[189] In 1878, Baedeker's *Egypt* touted the New Hotel as 'a large building with handsome rooms'.[190] By 1895, the guidebook warned visitors wishing to stay at the New Hotel that, although it was 'opposite the Ezbekiyeh garden, with terrace', it was 'also in a busy neighborhood'.[191] A hotel called the Continental had been in Cairo since 1889, in a converted mansion on Qasr al-Nil Street, not far from the Azbakeya. In Baedeker's 1895 edition, the Continental was described as being 'in a quiet situation in the new quarter of Ismailiya, newly added to and pleasantly fitted up, with terrace, gardens, and separate suites for families, fashionable'.[192] It seemed to be ideal for long-term visitors, and was in a prime, if not completely central, location.

By 1898, the New Hotel had been rebranded the Grand Hotel, and then bought by George Nungovich, the hotelier who had just lost his lease on the building housing the Continental. Not wanting to lose the business along with the building, Nungovich moved the Continental, as it were, to the old New Hotel. He renovated the building, adding a fourth floor and new wings, and renamed it the Grand Continental.[193] It quickly became the most popular hotel in Cairo, second only, of course, to Shepheard's.[194] In 1902, Baedeker's advised visitors that the Continental had 300 rooms, a terrace, and restaurant (Figure 2.7).[195] However, since the hotel was in the same spot that the New Hotel had occupied and it was in the same building as the Grand Hotel, the location was still pretty noisy and busy. The Continental was seen as a family hotel, one where people could stay for a whole season and not be bothered with partygoers or balls, as one may be at Shepheard's. Archaeologists like Breasted and Newberry saw its quieter clientele as a benefit for their peaceful discussions on the terrace, even

Figure 2.7 Stereoscope of the Grand Continental, c. 1900.

though the neighbourhood was louder. Harold Jones had also enjoyed the terrace and used it to run into friends – not totally unexpectedly. The terrace of the Grand Continental was not unlike that of Shepheard's. It was a central truth spot in the intellectual landscape for Carter, Jones, Breasted, and other Egyptologists who frequented the busy street-side tables.

By 1902, Howard Carter had worked his way up from Petrie Pup (trainee) and EEF archaeologist to Chief Inspector of Upper Egypt, living in Luxor. While in Luxor, he met most of the wealthy visitors who came through on their dahabeahs and made some important connections. He worked for Davis and Andrews in 1902 and, by 1904, he had been moved to the Inspectorate of Lower Egypt, based in Saqqara, meaning he was living near Cairo and would visit the city regularly.[196] He would visit Mena House, eat dinner at Shepheard's with Newberry or with Davis and Andrews, but he preferred the Continental. After a scandal at Saqqara in 1905, in which Carter took the side of the Egyptian workers and not a group of belligerent French tourists, he was relieved of his duties as Inspector and was generally free to paint as he liked.[197] He was moved to the Delta town of Tanta to be an inspector there, but he quickly bored of it, because he had gotten used to the 'high society and learned society' of the urban centres.[198] By the fall of 1906, Carter was staying at the Continental, which his biographer T. G. H. James noted was a place Carter 'could hardly have afforded at the time unless he was the guest of some affluent visitor whom he was accompanying'.[199] Using his connections to wealthy patrons, through Davis and Andrews and others, Carter began to work with George Herbert, the Fifth Earl of Carnarvon. Known simply as Carnarvon, he stayed at the Continental regularly when he was in Cairo, and it was customary for patrons to pay for their artists and excavators to stay in the nicer hotels near them. The pair worked closely together beginning in Luxor in 1907 and they would continue their partnership up to, and beyond, Carnarvon's death in 1923.[200]

During the First World War, the Continental and the Savoy, its sister hotel also owned by the Nungovich hotel company, along with a number of other hotels and residences, were taken over by the British Army for headquarters, barracks, and offices. Carter continued to stay at the Continental throughout the war while visiting Cairo, even as he excavated mainly in Luxor. As he had become accustomed to doing, Carter took advantage of the other guests in the hotel by using their expertise to help in his own projects. In the summer of 1915, he wrote to Albert Lythgoe, of the Metropolitan Museum of Art (MMA) in New York, to ask if he could borrow some time and the proficiency of the English archaeologist Hugh G. Evelyn-White. Evelyn-White, who had begun working with the MMA expedition in 1909, was serving in the British military in Egypt during the War and staying at the Continental.[201] Carter wrote to Lythgoe and asked: 'White is here at the Continental working on his publication on the "Sayings of Christ". I wonder if you will mind my asking him to help me later on the Classical stuff re Valley of the Kings.'[202] Lythgoe wrote to Carter saying he did not mind, and the two worked together on some of Carter's ideas about the Valley.[203] Carter was very likely rubbing shoulders with other soldiers and other officials who came through, as well, but it is hard to say this definitively.

Plenty of soldiers stayed at the Continental and it was not just Carter who was benefitting from their collaboration with Egyptologists. General Allenby, who had become close friends with Breasted, stayed at the Continental during the War, as did T. E. Lawrence, also known as Lawrence of Arabia.[204] In fact, at the start of the war, Lawrence had more in common with Breasted than he did with Allenby. An archaeologist in his own right, Lawrence had begun his career in 1910 at the ancient Hittite site of Carchemish under Leonard Woolley. They worked together to gather intelligence for a particularly significant site report, published by the Palestine Exploration Fund in 1914. *The Wilderness of Zin* was not only an important publication for archaeologists, but its detailed geography and cultural discussion also made it a crucial source for British military intelligence throughout the War.[205] Lawrence, most well known for his role in the Arab Revolt in 1916, continued to gather intelligence throughout the War for the British, as did other archaeologists such as Gertrude Bell and David Hogarth.[206]

After the First World War, the Continental was one of the few hotels (along with Shepheard's) that opened up for tourist business immediately. It was through contacts at the Continental that Allenby continued to rely on archaeologists gathering data, centring James Breasted in the British Mandate of the Middle East. On his way home from Cairo in February of 1920, Breasted was allowed to read Lawrence's report to Allenby about his own experiences and thoughts on the situation in Syria. Breasted recounted to his family that 'The French are so insanely jealous of Lawrence's power and influence among the Arabs, that the British have not published Lawrence's report for fear of offending the French. It is a pity, for it is an extraordinary document.'[207] That the British would let Breasted read this document would be a little unexpected, unless Breasted had particular clearances from Allenby that we do not know about.[208] After meeting frequently at the hotel, it is clear that The Continental was a place of international intrigue throughout the War and after. For both war and archaeological operations, the hustle and bustle of the Continental meant that Western powers needed to maximise the impact of their time there. Even though tourists could stay there, when Breasted arrived at the Continental in November of 1919, he remarked to his family that 'the country is full of British troops and at Shepheard's and here (the only two hotels that are open), one sees almost nothing but khaki on the terrace and in the dining room'.[209] There were a number of important meetings among Egyptologists and political figures, and sometimes they were one and the same. Not only was the Continental in a central part of Cairo, but it also became a centre of the social, political, and professional scene after the War.

The Continental also played a major role in the most sensational Egyptian find of the twentieth century: the uncovering of the tomb of Tutankhamun. Most of the story of the uncovering of the tomb and the international issues it enabled is below in Chapter 4, but some parts of the story took place in Cairo. Once Carnarvon arrived in Luxor, the tomb doors were officially breeched on 26 November 1922; Carnarvon then had to return to England to take care of some matters there. Carter travelled back with his patron as far as Cairo, and a letter on 16 December shows Carnarvon staying at the Continental Hotel.[210] Carter would stay in Egypt working on the tomb and Carnarvon would return in early 1923 to continue supervising and helping to clear the tomb of its objects. From February to March, the work continued briskly, with a number of Egyptologists famously dedicating their own field seasons to aiding this work. In the middle of March 1923, Carnarvon got a mosquito bite on his cheek, which then became infected. He refused to rest, even though he was not feeling at all well. The bite likely caused septicemia, a bacterial blood infection that could not be cured at the time. Carnarvon died in the Continental on 5 April 1923. Alan Gardiner wrote about Carnarvon and his untimely death:

> He might, perhaps have recovered from the mosquito bite which he got in Luxor if he had taken better care of himself. Disregarding the doctor's advice he came down to Cairo and invited me to dine with him at the Mohammed Ali Club. He expressed himself very tired and despondent but insisted on going to a film. There he said that his face was hurting him and I begged him to go back to his hotel, the Continental. But no, he would see the film to a finish and he was never out of doors again.[211]

This event fanned the flames of the theory that a mummy's curse killed the man who found King Tutankhamun's virtually undisturbed tomb. In reality, it was a bacterial infection that ran its deadly course before the advent of reliable antibiotics. Carnarvon's death was an unlucky circumstance that also weakened the attempts that the struggling Carter made to maintain British control over the tomb (see Chapter 4). Interestingly, in September of 1925, the hotel manager at the Continental, wanting to honour Carnarvon's memory, placed Carter in the same room that Carnarvon had died in.[212]

The Continental remained a favourite of visitors, including Egyptologists and their patrons, throughout the first half of the twentieth century. It was in business until at least the 1950s, but it is not clear when it stopped being a hotel. Since at least the 1980s there had been an inoculation clinic for

Egyptians travelling to the rest of Africa. Humphreys paints a picture of the long, drawn-out end of a lively and well-loved hotel: 'By the 1980s, the only park was the car park where the opera used to stand; half the Ezbekiyya had been concreted over and the rest was a dusty wasteland. Tourists now preferred to stay beside the Nile, where the river breezes made the air more breathable.'[213] The building was finally demolished in February 2018, and it is unclear what will be built in its place.

Conclusions

This chapter of the book is long because there were more hotels in Cairo and its environs than in the rest of Egypt in this period. Cairo was a city that welcomed guests from all over the world, and their purposes varied as widely as their backgrounds. Archaeologists and patrons used the city of Cairo and its hotels as a staging ground to begin trips up the Nile, to go further out into the field, and to prepare them for their futures. There were more hotels in Cairo, and more people came through Cairo on their way to other places in the world, but most archaeological activity that we recognise as such happened outside of the city. However, in Cairo, hotels were truth spots in which Egyptologists met, became professionals, and shared in others' work. They were peaks in the intellectual landscape in which archaeologists found patrons, did research, met with friends, got jobs, and formed their ephemeral scientific networks. Whether it was Budge learning about the Cairo antiquities market, or Jones finding a job, these hotels were the places that significant scientific work was done. Their time in the hotels helped to prepare archaeologists for all of these physical and academic journeys by providing them with experience and new tools for their excavation seasons. Some of the connections lasted decades, some only a few months, but they were foundational connections that set many Egyptologists up for successful careers. There are countless more we do not know about because they did not make it into the published record or survive in archival collections.

As they left Cairo on their steamers, dahabeahs, or trains, these different types of journeys and the destinations to which they would go would lead to different ways in which they engaged with each other and the ideas they had begun to gather in Cairo.

Notes

1 Baedeker's *Egypt* (1878), 226.
2 Mary Brodrick, *A handbook for travellers in Lower and Upper Egypt* (London: John Murray, 1900), 229.

3 E. A. Wallis Budge, *By Nile and Tigris: A narrative of journeys in Egypt and Mesopotamia on behalf of the British Museum between the years 1886 and 1913* (London: John Murray, 1920), 80.
4 Charles Breasted, *Pioneer to the past: The story of James Henry Breasted archaeologist* (Chicago: The Oriental Institute, 1943), 62.
5 Hilda Petrie, 22 December 1897, in Drower, ed. *Letters from the desert*, 122–3.
6 Brodrick, *Handbook for travellers*, 229.
7 Gieryn, *Truth spots*.
8 See Colla, *Conflicted antiquities*; Reid, *Whose pharaohs?*; Reid, *Contesting antiquity*.
9 It is important to note that Egyptians were largely excluded from scholarly training in Egyptology, both in and outside Egypt, until the early years of the twentieth century, with the notable exception of the School of Ancient Language, supported by ʻAli Mubarak and led by Heinrich Brugsch in the 1870s. It failed shortly after being founded (see Colla, *Conflicted antiquities*, 142–65; Reid, *Whose pharaohs*, 116–18, 120–4). Then, even with proper training, they continued to be excluded from the Western study of Egyptology until around 1935 (Reid, *Contesting antiquity*, 109–33, esp. 117–21).
10 See especially Reid, *Whose pharaohs?*, 201–4; Reid, *Contesting antiquity*, 29–33.
11 See, for example, James Aldridge, *Cairo* (Boston: Little, Brown and Company, 1969); Mohamed Elshahed, *Cairo since 1900: An architectural guide* (Cairo: AUC Press, 2019); Deborah Manley, *A Cairo anthology: Two hundred years of travel writing* (Cairo: AUC Press, 2013); André Raymond, transl. Willard Wood, *Cairo* (Cambridge, MA: Harvard University Press, 2000); Max Rodenbeck, *Cairo: The city victorious* (New York: Vintage Books, 1998); David Sims, *Understanding Cairo: The logic of a city out of control* (Cairo: AUC Press, 2010).
12 H. R. Hall, ed. *Murray's, A handbook for Egypt and the Sudan* (London: John Murray, 1907), 89.
13 Baedeker's *Egypt* (1885), 231.
14 Baedeker's *Egypt* (1895), 27. For more thorough histories of Thomas Cook, please see Piers Brendon, *Thomas Cook: 150 years of popular tourism* (London: Secker & Warburg, 1991); also, Andrew Humphreys' two books *Grand hotels* and *On the Nile*.
15 Baedeker's *Egypt* (1885), 231.
16 Baedeker's *Egypt* (1895), 28.
17 Baedeker's *Egypt* (1885), 239–40.
18 This is true for Baedeker's, Murray's, and Cook's travel guides.
19 Brendon, *Thomas Cook*, 120–1.
20 Baedeker's *Egypt* (1914), 35.
21 Baedeker's *Egypt* (1895), 34.
22 Eustace Alfred Reynolds-Ball, *Cairo of to-day; a practical guide to Cairo and the Nile* (London: A and C Black, 1916), 25–35.
23 In the next chapter, I will detail some of these memoirs travellers may have used.
24 Aldridge, *Cairo*, 196.

25 *Ibid.*, 178; Raymond, *Cairo*, 183.
26 Aldridge, *Cairo*, 159.
27 Vivant Denon, *Voyage dans la Basse et la Haute Egypte pendant les campagnes du General Bonaparte* (Paris: P. Didot, 1802) contains beautiful images, one of which is entitled 'Cairo viewed from Ezbekiya square during flooding of the Nile, 1802' engraved by Louis-Pierre Baltard.
28 Aldridge, *Cairo*, 188.
29 See, for example, Isolde Lehnert, '"Let's have a beer at Gorff's!"' in *Journeys erased by time: The rediscovered footprints of travellers in Egypt and the Near East*, ed. Neil Cooke (Oxford: Archaeopress, 2019), 115–32.
30 *Cook's tourists' handbook for Egypt, the Nile, and the Desert* (London: Thomas Cook & Son, 1897), 101–2.
31 J. Lane MSS 1, 1 December 1873, Journal 1, Jenny Lane Collection, Griffith Institute of Egyptology, Oxford University.
32 Deborah Manley, ed. *Women travellers on the Nile* (Cairo: AUC Press, 2016), 28.
33 Mary J. Holmes, 'Street Life in Egypt', *St. Louis Post Dispatch* (9 February 1890): 24.
34 *Cook's tourists' handbook for Egypt* (1897), 102.
35 Mitchell, *Colonising Egypt*, 1–33.
36 *Ibid.*
37 Very often in these exhibition recreations, the architects would import Egyptian donkeys and even make the paint on the buildings dirty (Mitchell, *Colonising Egypt*, 1). Also, Sadiah Qureshi, *Peoples on parade: Exhibitions, empire, and anthropology in nineteenth century Britain* (Chicago: University of Chicago Press, 2011), esp. 235–56.
38 JHB to his family, 4 November 1894, JHB Papers, Box 4, JHB Correspondence.
39 Mitchell, *Colonising Egypt*, 22.
40 Gehan Selim, *Unfinished places: The politics of (re)making Cairo's old quarters* (London: Routledge, 2017), 65.
41 See Stevenson, *Scattered finds*, 259–60.
42 Humphreys, *Grand hotels*, 101.
43 Baedeker's *Egypt* (1885), 233.
44 William Thackeray, quoted in Michael Haag, ed. *An Alexandria anthology: Travel writing through the centuries* (Cairo: AUC Press, 2014), 58.
45 Twain, *Innocents abroad*, 396–7.
46 Humphreys, *Grand hotels*, 104.
47 Petrie, 'A Digger's Life', 440–1.
48 Humphreys, *Grand hotels*, 107; Auguste Wild, *Mixed grill in Cairo: Experiences of an international hotelier* (London: Sydenham & Co., 1954), 57.
49 Baedeker's *Egypt* (1895), 28.
50 *The Egyptian Gazette*, 20 January 1900.
51 Humphreys, *Grand hotels*, 111; Pierre Loti, *Egypt (La Mort de Philae)* (New York: Duffield & Co., 1910), 7–8.
52 *The Egyptian Gazette*, 10 November 1894. See Table 0.1 for conversion. This is equivalent to £17.50 or $31.50 today.

53 Baedeker's *Egypt* (1895), 30.
54 *Ibid.*
55 Baedeker's *Egypt* (1902), 112–13; original emphasis.
56 *Ibid.*, 112.
57 Manley, *Women travellers*, 19–20.
58 Edwards, *A thousand miles up the Nile*, 31.
59 JHB to his family, 4 November 1894, JHB Papers, Box 4, JHB Correspondence.
60 Hilda Petrie, 22 December 1897, in Drower, ed., *Letters from the desert*, 123.
61 George Reisner, the American Egyptologist, made his career at Giza from 1902–08. He likely frequented the hotel and possibly stayed there from time to time while working.
62 James, *Howard Carter*, 124.
63 In 1916, Reynolds-Ball told readers of *Cairo of to-day* that 'Helouan is apt to be dull and depressing', owing to the fact it was so far removed from any sights of the city (218).
64 Reynolds-Ball, *Cairo of to-day*, 217–18.
65 Norman de Garis Davies to Gruber, 9 February 1905, EES V.h.38; Norman de Garis Davies to Gruber, 4 March 1905, V.h.45.
66 Reynolds-Ball, *Cairo of to-day*, 53.
67 Morgan and Eddisford, 'Dig Houses', 169–93; Carruthers, 'Credibility, civility, and the archaeological dig house', 255–76. I discuss dig houses below, in Chapters 3 and 4.
68 Petrie MSS 1.3.1–50, 22 November 1883, Petrie Journal 1883–84, Griffith Institute of Egyptology, Oxford University.
69 James, *Howard Carter*, 38.
70 Percy E. Newberry, 'Howard Carter', *Journal of Egyptian Archaeology* 25:1 (1939): 67.
71 Katherin Blouin, *Triangular landscapes: Environment, society, and the state in the Nile Delta under Roman rule* (Oxford: Oxford University Press, 2014), 47; Percy E. Newberry, 'B. The Archaeological Survey of Egypt: Mr. Newberry's Work, 1892–93', *Egypt Exploration Fund, Archaeological Report 1892–1893* (London: Kegan Paul, 1893), 9.
72 James, *Howard Carter*, 51.
73 Guthrie Roger to Percy Newberry, 3 February 1893, NEWB2/624, Griffith Institute of Egyptology, Oxford University.
74 *Ibid.*
75 Guthrie Roger to Emily Paterson, 19 April 1893, EES VI.b.10–12.
76 Humphreys, *Grand hotels*, 50.
77 Sir John Gardner Wilkinson, *Hand-book for travellers in Egypt; including descriptions of the course of the Nile to the second cataract, Alexandria, Cairo, the pyramids, and Thebes, the overland transit to India, the peninsula of Mount Sinai, the oases, &c. Being a new edition, corrected and condensed, of 'Modern Egypt and Thebes* (London: John Murray, 1847), 117–18.
78 Baedeker's *Egypt* (1885), 231.
79 Petrie 'A Digger's Life'.

80 http://grandhotelsegypt.com/?p=1118 (accessed 15 August 2018).
81 Baedeker's *Egypt* (1908), 31.
82 http://grandhotelsegypt.com/?p=1118 (accessed 15 August 2018).
83 *Ibid.*
84 Harold Jones to his parents, 19 December 1905, Garstang Museum Archives, FC/1/4 Xeroxes of Ernest Harold Jones.
85 Archibald D. Walker, *Egypt as a health-resort* (London: Churchill, 1873); Richard E. Morris, 'The Victorian "Change of Air" as medical and social construction', *Journal of Tourism History* 10:1 (2018): 49–65.
86 See, among others, Toby Wilkinson and Julian Platt, *Aristocrats and archaeologists: An Edwardian journey on the Nile* (Cairo: AUC Press, 2017).
87 Walker, *Egypt as a health-resort*; Morris, 'The Victorian "Change of Air"'.
88 Wilkinson and Platt, *Aristocrats and archaeologists.*
89 *Ibid.*, 126.
90 C. Delany, *A son to Luxor's sand: A commemorative exhibition of Egyptian art from the collections of the British Museum and Carmarthen Museum* (Dyfed City Council, 1986).
91 *Ibid.*, 5.
92 Harold Jones to his family, nd (1903/4?), Garstang Museum Archives FC/1/4 Xeroxes of Ernest Harold Jones.
93 Harold Jones to his parents, 19 December 1905, Garstang Museum Archives FC/1/4 Xeroxes of Ernest Harold Jones.
94 This is not to say that photography was not done on site. Flinders Petrie used early photographic methods. Harry Burton's photographs of the excavation of the tomb of Tutankhamun are also famous (see, for example, Christina Riggs, *Photographing Tutankhamun: Archaeology, ancient Egypt, and the archive* [London: Routledge, 2019]). However, as Breasted began, the Epigraphic Survey at the University of Chicago still follows what is called the Chicago House method, which involves a mix of photography and hand drawing for greater detail (https://oi.uchicago.edu/research/projects/epi/chicago-house-method, accessed 22 September 2020).
95 C. Delany, *A son to Luxor's sand*, 6.
96 Harold Jones to his family, 2 March 1906, Garstang Museum Archives FC/1/4 Xeroxes of Ernest Harold Jones.
97 Harold Jones to his family, 30 May 1906, Garstang Museum Archives FC/1/4 Xeroxes of Ernest Harold Jones.
98 *Ibid.*
99 Harold Jones to his family, 2 March 1906 and 27 May 1906, Garstang Museum Archives FC/1/4 Xeroxes of Ernest Harold Jones; £300 would be the equivalent of £25,000 or $34,863 today.
100 These appeared in Theodore M. Davis, *The tomb of Siphtah; The Monkey tomb and the Gold tomb* (London: Archibald Constable & Co., 1908) and Theodore M. Davis, *The tomb of Queen Tiyi* (London: Duckworth, 1910).
101 See Abt, *American Egyptologist*, 6–19.
102 See Sheppard 'Trying desperately to make myself an Egyptologist', 174–87.

103 C. Breasted, *Pioneer to the past*, 51. Published between the years 1926 and 1961, the dictionary has become essential to anyone studying ancient Egyptian scripts. Adolf Erman, Hermann Grapow, eds, *Wörterbuch der Aegyptischen Sprache* (Berlin: Akademie-Verlag, 1926–61); it is usually abbreviated *Wb*.
104 Quoted in C. Breasted, *Pioneer to the past*, 62.
105 Humphreys, *Grand hotels*, 50.
106 William Howard Russell, *My diary in India, in the year 1858–9* (London: Routledge, Warne, 1860), 28.
107 Baedeker's *Egypt* (1887), 229.
108 See Table 0.1 for currency conversion. This would have been between 7s 11d–9s 10d per day at the time, or £44.18–55.41 and $66.75–$83.70.
109 Petrie, 'A Digger's Life'.
110 Wallis Budge, *By Nile and Tigris*, 83.
111 Wallis Budge, *By Nile and Tigris*.
112 W. Benson Harer, Jr. 'The Drexel Collection: From Egypt to the Diaspora,' in Sue H. D'Auria, ed. *Servant of Mut: Studies in honor of Richard A. Fazzini* (Leiden: Brill, 2008), 111–19; Heicke Schmidt, 'The Notorious Emil Brugsch: "It is said that Brugsch Bey would sell the whole museum"', in Neil Cooke, ed. *Journeys erased by time: The rediscovered footprints of travellers in Egypt and the Near East* (Durham and Oxford: ASTENE and Archaeopress, 2019), 81–99.
113 http://grandhotelsegypt.com/?p=929 (accessed 15 August 2018).
114 Wild, *Mixed grill*.
115 *Ibid*., 57.
116 Baedeker's *Egypt* (1895), 27.
117 Baedeker's *Egypt* (1902), 24. See Table 0.1 for conversion: 50 piastres was equivalent to 10s 3d at the time, or $2.50. This corresponds to the purchasing power of about £44.60 or $77.00 today.
118 Sheppard 'Trying desperately to make myself an Egyptologist'.
119 C. Breasted, *Pioneer to the past*, 140–2.
120 *Ibid*., 146.
121 *Ibid*., 147; Abt, *American Egyptologist*, 126–53.
122 C. Breasted, *Pioneer to the past*, 147.
123 Gaston Maspero to JHB, 16 October 1905, JHB Office Files, 1905, JHB Correspondence.
124 Andrews' Diary, 28 November 1906.
125 http://grandhotelsegypt.com/?p=929 (accessed 15 August 2018).
126 For example, Nina Nelson, *Shepheard's Hotel* (London: Barrie and Rockliff, 1960) and others cited throughout this chapter.
127 Lanver Mak, *The British in Egypt: Community, crime and crises 1882–1922* (London: I. B. Tauris, 2012), 83–117.
128 Michael Bird, *Samuel Shepheard of Cairo: A portrait* (London: Michael Joseph, 1957), 49; Humphreys, *Grand hotels*, 76.
129 Bird, *Samuel Shepheard*, 34–5.
130 *Ibid*.

131 Humphreys, *Grand hotels*, 80.
132 Edwards, *A thousand miles up the Nile*, 2.
133 See Eleanor Dobson, 'A tomb with a view: Supernatural experiences in the late nineteenth century's Egyptian hotels', in *Anglo-American travellers and the hotel experience in nineteenth-century literature: Nation, hospitality, and travel writing*, eds Monika M. Elbert and Susanne Schmid (London: Routledge, 2017), 89–105.
134 Anthony Trollope, 'An unprotected female at the pyramids', in *Tales of all countries* (London: Chapman and Hall, 1867), 141.
135 *The Camels are Coming* (1934).
136 Peters, *Crocodile on the sandbank*, 31–2.
137 Humphreys, *Grand hotels*, 80–2.
138 Humphreys, *Grand hotels*, 81–2; Morris Bierbrier, *Who was who in Egyptology* (London: Egypt Exploration Society, 2012), 124.
139 *Ibid.*
140 Tarek Ibrahim, *Shepheard's of Cairo: The birth of the Oriental grand hotel* (Wiesbaden: Reichert Verlag: 2019).
141 Ibrahim, *Shepheard's of Cairo*, 28–9.
142 Humphreys, *Grand hotels*, 81–2.
143 Ibrahim, *Shepheard's of Cairo*, 32; Humphreys, *Grand hotels*, 82.
144 *Ibid.*, 85.
145 Nelson, *Shepheard's Hotel*, 57.
146 Ibrahim, *Shepheard's of Cairo*, 29.
147 W. M. Flinders Petrie, Memoranda to the Committee, 1 November 1904, EES.V.h.20.
148 John Cook to Emily Paterson, 17 November 1893, EES.XII.f.16
149 W. E. Kingsford for Thomas Cook to Emily Paterson, 12 November 1894, EES.XId14 and EES.XId28.
150 Felix Marschner to John Ward, 22 August 1904, EES.V.h.68.
151 Holmes, 'Street Life in Egypt'.
152 Humphreys, *Grand hotels*, 88.
153 John M. Adams, *The millionaire and the mummies: Theodore Davis's Gilded Age in the Valley of the Kings* (New York: St Martin's Press, 2013), 6.
154 *Ibid.*
155 The *Emma B. Andrews Diary Project*, spearheaded by Sarah Ketchley at the University of Washington, is in the process of transcribing these diaries, among other projects associated with it. I was able to view the typewritten copies at the MMA, through the kind permission of Catharine Roehrig. It is the MMA copies I cite from unless otherwise noted.
156 Andrews' Diary, 12 December 1889.
157 *Ibid.*
158 *Ibid.*
159 James Goode, *Negotiating for the past: Archaeology, nationalism, and diplomacy in the Middle East, 1919–1941* (Austin: University of Texas Press, 2007), 99–100.

160 See Kathleen Sheppard, 'On His Majesty's Secret Service: James Henry Breasted, accidental spy', *Journal of the Society for the Study of Egyptian Antiquities* 44 (2017–18): 251–72.
161 See Table 0.1 for conversion. This amount corresponds to approximately $306,000 per year for 5 years in 2021 USD.
162 The University of Chicago has recently transcribed and published an online volume of all of these letters. I have been to the Oriental Institute to view these letters, but I will be using the publication for quoting Breasted's words. John A. Larson, ed. *Letters from James Henry Breasted to his family, August 1919–July 1920: Letters home during the Oriental Institute's first expedition to the Middle East*. Oriental Institute Digital Archives, No. 1 (Chicago, 2010). Larson, ed. *Letters from JHB*, 3 November 1919.
163 Larson, ed. *Letters from JHB*, 6 November 1919.
164 See Abt, *American Egyptologist*, esp. 256–65. Larson, ed. *Letters from JHB*, 5 November 1919.
165 *Ibid.*
166 Larson, ed. *Letters from JHB*, 6 November 1919.
167 Bierbrier, *Who was who in Egyptology*, 216.
168 Larson, ed. *Letters from JHB*, 6 November 1919.
169 *Ibid.*
170 *Ibid.* See Table 0.1 for conversion. This is equivalent to about $20 today.
171 Larson, ed. *Letters from JHB*, 10 November 1919.
172 *Ibid.*; James Breasted, *A history of Egypt from the earliest times to the Persian Conquest*, 2nd edn (London: Hodder and Stoughton, 1927). There are almost 200 illustrations in this second volume, but few are attributed, and those that are, are not attributed to Brunton.
173 Winifred Brunton, *Kings and queens of Ancient Egypt. Portraits by Winifred Brunton. History by eminent Egyptologists, Etc.* (London: Hodder and Stoughton, 1924). The contents were organised around Brunton's paintings, with chapters by T. E. Peet, Margaret Murray, and more.
174 Larson, ed. *Letters from JHB*, 10 November 1919.
175 Larson, ed. *Letters from JHB*, 16 November 1919.
176 *Ibid.*
177 Larson, ed. *Letters from JHB*, 2 November 1919.
178 Larson, ed. *Letters from JHB*, 30 November 1919.
179 The issues that Allenby and the British were dealing with at this time were important in terms of the colonial presence of the British in Egypt. For a fuller treatment on Breasted's role in particular, see Sheppard 'On His Majesty's Secret Service'.
180 Larson, ed. *Letters from JHB*, 30 November 1919.
181 *Ibid.*
182 *Ibid.*
183 *Ibid.*
184 John A. Larson, 'Introduction', *Letters from James Henry Breasted to His Family, August 1919–July 1920*, ed. John A. Larson (Oriental Institute Digital Archives, Number 1, 2010), 11.

185 James Barr, *A Line in the sand: Britain, France and the struggle for the mastery of the Middle East* (London: Simon & Schuster, 2011). Also useful for this context are James Barr, *Setting the desert on fire: T. E. Lawrence and Britain's secret war in Arabia, 1916–1918* (New York: W. W. Norton & Co., 2008); James Barr, *Lords of the desert: The battle between the United States and Great Britain for supremacy in the modern Middle East* (New York: Basic Books, 2018).
186 Sheppard, 'On his Majesty's Secret Service'.
187 See Abt, *American Egyptologist*, 207–48; Larson, ed. *Letters from JHB*; Sheppard, 'On his Majesty's Secret Service'.
188 'Condemns Roosevelt Speech', *New York Times* (1 April 1910): 4; Adams, *Millionaire and the mummies*, 76–7.
189 Humphreys, *Grand hotels*, 122–3.
190 Baedeker's *Egypt* (1878), 229.
191 Baedeker's *Egypt* (1895), 27.
192 *Ibid.*
193 Humphreys, *Grand hotels*, 123.
194 *Ibid.*
195 Baedeker's *Egypt* (1902), 24.
196 James, *Howard Carter*, 112.
197 *Ibid.*, 131–7; H. V. F. Winstone, *Howard Carter and the discovery of the tomb of Tutankhamun*, revised edition (Manchester: Barzan, 2006), 88–95.
198 James, *Howard Carter*, 137.
199 *Ibid.*, 157.
200 For more about their work together, see Chapter 4.
201 Bierbrier, *Who was who in Egyptology*, 183.
202 James, *Howard Carter*, 205.
203 See Chapter 4, below.
204 Humphreys, *Grand hotels*, 124–5.
205 C. Leonard Woolley and T. E. Lawrence, *The wilderness of Zin* (London, 1914).
206 Janet Wallach, *Desert queen: The extraordinary life of Gertrude Bell adventurer, adviser to kings, ally of Lawrence of Arabia* (London: Phoenix/Orion Books Ltd, 1997), 115–18; Barr, *Setting the desert on fire*. It is not clear if Breasted knew Lawrence, but he knew and liked Bell very much.
207 Larson, ed. *Letters from JHB*, 27 February 1920.
208 See Barr, *A line in the sand*; Sheppard, 'On His Majesty's Secret Service'.
209 Larson, ed. *Letters from JHB*, 2 November 1919.
210 James, *Howard Carter*, 266–7.
211 James, *Howard Carter*, 297, from Alan Gardiner's *My working years* (Privately published, 1963), 40.
212 Winstone, *Howard Carter*, 268; Humphreys, *Grand hotels*, 129.
213 http://grandhotelsegypt.com/?p=2535 (accessed 15 August 2018).

3

Up the Nile: *l'esprit du Nil*

> A rapid raid into some of the nearest shops, for things remembered at the last moment – a breathless gathering up of innumerable parcels – a few hurried farewells on the steps of the hotel – and away we rattle as fast as a pair of rawboned greys can carry us ... And now at last all is ready ... and away we go, our huge sail filling as it takes the wind! Happy are the Nile travellers who start thus with a fair breeze on a brilliant afternoon.[1]

So began Amelia Edwards' first and only trip up the Nile on the rented dahabeah *Philae* in 1874, a journey that would create not only an Egyptologist of her, but also would be a catalyst to bring about the creation of Egyptology as a university-taught subject and the professionalisation of the discipline in the UK.[2]

The archaeologists and travellers from the previous chapters needed to go up the Nile for their work, and many did so for pleasure as well. The situation was usually the same for both archaeologists and tourists – once their preparation time was finished in a tourist-filled Cairo, they embarked on a trip up the Nile. In the late nineteenth and early twentieth century, some tourists on the fast track with Thomas Cook might take a train south to Luxor or Aswan, then turn around and take a steamboat back down the Nile, north to Cairo. Others would take a steamer up and back; still others – those with the most time and money – would rent a houseboat, a dahabeah, for the season; some people had their own dahabeahs built so they could come back each year. Many times, they tried to get to Luxor or Aswan as quickly as they could, sailing with the wind, only to turn around and leisurely sail with the current back to Cairo, eventually. Along the way, they would stop at various towns and archaeological sites, sometimes staying several days or weeks at particular places. This chapter will detail the travel options in order of speed and cost, and will of course discuss the ways in which archaeologists interacted with one another on trips up the Nile and how that activity would impact Egyptology. Archaeologists spent a great deal of time on dahabeahs as they moved about the country, so this chapter will spend the most time discussing work and life on a dahabeah in this period.

These ephemeral travel spaces were in use by archaeologists for varying amounts of time, depending on their conveyance of choice or necessity. As Christina Riggs and others argue, where archaeologists could travel, by river or rail, influenced where they could work and for how long.[3] This chapter will further interrogate these travel systems and argue that in terms of understanding the work archaeologists would do once they arrived on site, modes of travel reveal their methods of work and their levels of productivity. Depending on the amount of time spent in their respective modes of travel, archaeologists imbued these spaces with varying levels of influence in the creation of archaeological knowledge or activity in the discipline. Taking a train meant a hurried commute to work with little time to talk, think, or write on the way. Taking a tourist steamer or a dahabeah meant that archaeologists could see more than just their own site of work through touring stops, could meet others who were touring or working, and had time to establish themselves as experts in the discipline. Dahabeahs, while travelling on the Nile, were semi-private spaces that operated both as a home away from home and as spaces where patrons and their excavators could give and receive valuable archaeological training. Once moored at their final location, the boats would operate in the same way as a semi-public hotel space or dig house would, in that dahabeah owners or renters had the power to allow or deny visitors into their private domestic space. Further, archaeologists worked and met in the common areas on board, often storing excavation finds there which made it necessary for outsiders to come aboard into their public workspace.[4] In order to understand these methods of transport – especially steamers and dahabeahs – as sites of knowledge creation and the settings on the intellectual landscape that our Egyptologists were building, using, and working within, we must really spend some time on the river. The boats, because of the ways their passengers used them, became scientific institutions themselves.

Up the Nile: train, steamer, dahabeah

As travellers prepared to leave Cairo for sites and sights further south, they had to choose how to travel. By the turn of the twentieth century, those with enough time and money had quite a few options. Guidebooks such as Baedeker's, Murray's, and those published by Thomas Cook & Son gave advice to travellers on how to travel from Cairo to the rest of Egypt, and the details in those books was useful for archaeologists and other travellers as they made their way up the Nile. In this section, I will discuss travel first by train, then steamer, then dahabeah, and travellers will share their stories along the way.

Trains

While the rail line from Alexandria to Cairo was complete by 1865, the rail line through from Cairo to Luxor was not finished until 1898 (Figure 3.1). Before then, travellers had to parse together train journeys with steamers or dahabeahs if they wanted to go all the way to Luxor or further south. No doubt the train was the fastest and cheapest way to get through the country before airplanes, once the lines were completed, but they were said by many Western travellers to be difficult and miserable to travel by because of the heat and dust.[5]

Starting in December of 1888, Cook's offered what they called a 'cheap express service' which was not a guided tour of Egypt like those that steamer passengers enjoyed, but instead a train–steamer combination service designed to take people to Luxor and Aswan much more quickly than by steamer alone.[6] The express service involved taking the train from Cairo to Asyut, about ten hours (235 miles) by rail and still below (or to the north of) Luxor. Once at Asyut, travellers could get on their pre-booked steamer or dahabeah, or find berth on a mail steamer if one was available. The itinerary would then take them to Belianeh for Abydos and other sites, Denderah, and then on to Thebes at Luxor. Travellers could visit sites between Asyut and Luxor along the way, if they chose. If those using this travel scheme left Cairo on a Tuesday, they could arrive in Luxor by Friday.[7] Train travellers could still see the pyramids on their way out of town and watch the river go by. Cook's offered a package deal for this trip, because they offered package deals for most of their trips. In 1888, an eleven-day trip to Luxor and back by train and steamer cost travellers £20, and a fifteen-day trip to Aswan and back £25; servants cost £10 less because they took second class.[8] These costs included train and steamer tickets, food on the steamer, and accommodation at Luxor for four nights at either Cook's Luxor or Karnak Hotels.[9] Later on, there were even shorter trips on the Nile, which would include taking a train from Cairo to Luxor or Aswan, then taking a Cook's steamer or mail steamer down the Nile at a not-particularly-leisurely pace. That trip would take about a week and could cost between £16 and £25.[10] All of this was inclusive of room, board, and excursions but exclusive of getting to Egypt to begin that journey.

By 1898, Baedeker's *Egypt* told travellers that the completion of the railway from Cairo to Luxor would open up Egypt to travellers to move between Luxor and Aswan more easily, and they were right. From Cairo to Aswan by rail was 517 miles, which would take about twenty-two hours, and travellers had to switch at Luxor to take the narrow gauge train from Luxor to Aswan. From Cairo to Luxor by rail was a distance of 417 miles, which would take sixteen hours. In fact, Baedeker's suggested that 'experienced travellers may hire a tent and a small canteen at Cairo, engage

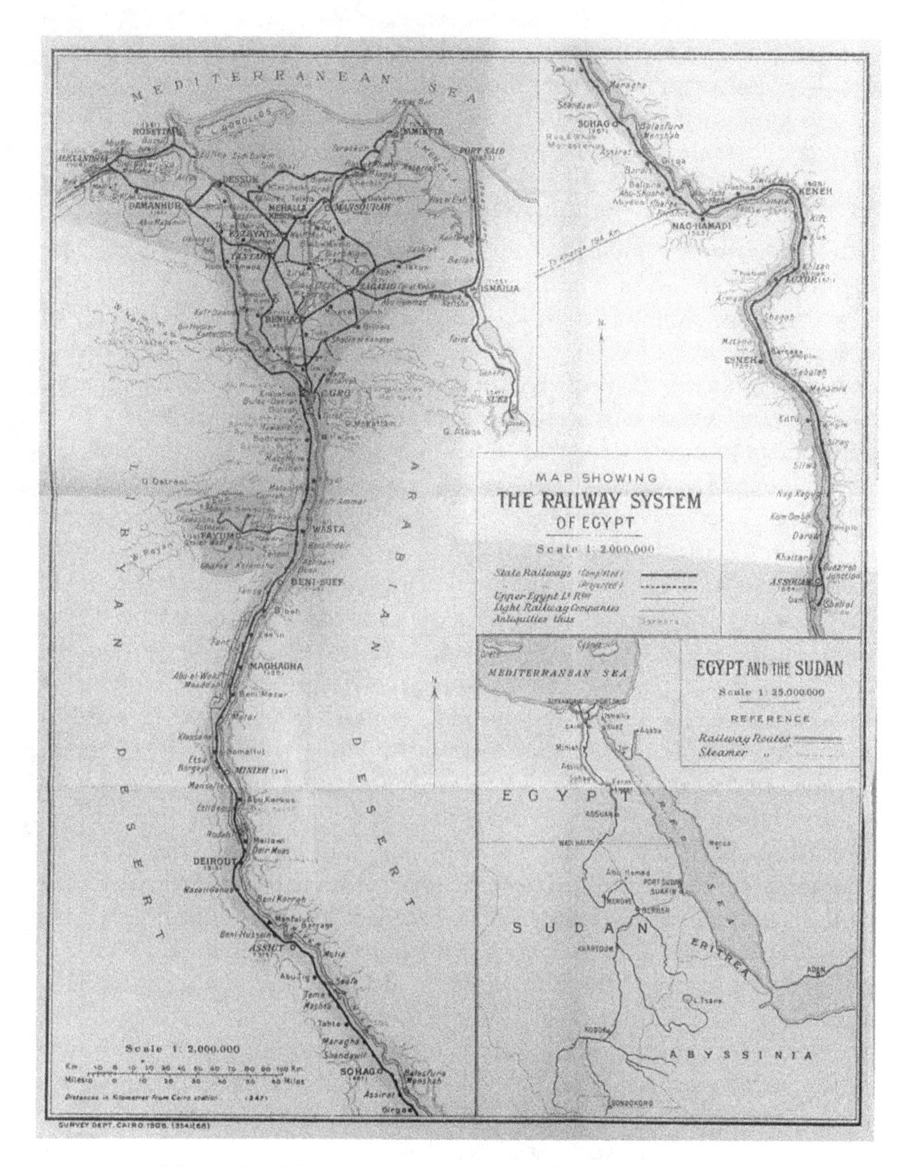

Figure 3.1 Map of Egyptian State Railway Lines c.1908.

a donkey-boy to act as cook, and proceed by train, stopping at the places they intend to visit'.[11] This would allow anyone who had the time and desire to see the sights relatively cheaply, but also at their own pace. It was a dahabeah-like option without the cost. I have not found any extant reports of anyone doing this, but Cook, ever in favour of the package-deal, suggested no such thing.[12]

Much like for travel up the Nile by boat, guidebooks contained sections detailing how train passengers should prepare themselves for the trip and how to book their passage. To travel by train, first class, was obviously the most comfortable for European travellers, but second class was acceptable if one wished 'to make a nearer acquaintance with the country and the people'.[13] This meant that second-class passengers would be travelling with Egyptians, eating Egyptian food, and experiencing a different train experience than those in the wholly European first class. When travelling by train to Luxor, tourists were told they 'should provide themselves with a supply of meat, bread, and wine, as no stoppage is made for dinner; eggs, bread, water (glass necessary) are offered for sale at the stations'.[14] Most importantly, as with everywhere they went in Egypt, travellers would be led by their guidebooks to watch for sights out of their windows on their journey. On the way out of Cairo, the train would immediately cross to the West side of the Nile, and the right side of the train car had views of the pyramid field from Giza to Dashur.

Upon arrival at El Wasta, fifty-seven miles south of Cairo, travellers could change for the Fayum train, stop to see sites in that area, or remain on the train. Baedeker's guidebook would ride with travellers all the way up the Nile, stopping at every rail station along the way, describing the sites that could be seen at each location if passengers chose to get off the train. Most would stay on the train if they were on the longer journey to Luxor or Aswan, but the option was available; 345 miles from Cairo, the train crossed an iron bridge going from the West to the East Bank of the river at Nag Hamadeh, which had been the end of the line until 1897, and then travelled another seventy-two miles to Luxor. The station in Luxor was in the southeast of the town, where it remains, and passengers would change here for the narrow-gauge rail line to Aswan. For the purposes of this book, we are not going to Aswan, although many travellers, including archaeologists, certainly went there – and further south.

The train journey was relatively quick and the corresponding narrative in the guidebooks was brief. The relative affordability of the train journey, in time and expense, allowed for 'persons of less ample means' to see some sights on the way to Luxor as well as sights further south.[15] Many archaeologists, especially those on a budget like Petrie, and Breasted early in his career, would take trains to get to their sites quickly and cheaply. In fact, the Egypt Exploration Fund had an agreement with Thomas Cook that those working for the Fund would be eligible for half-rate tickets.[16] It is unclear if the rate reduction was meant for all forms of travel, including boats, but one can assume it meant for train tickets at the very least. Egyptologists also combined modes of travel – many who would take a train would also use steamers and dahabeahs on those same trips.

Steamers

Thomas Cook had only moved permanent offices into Cairo in 1873 and 'in that year instituted the first regular service of steamers between Cairo and Aswan'.[17] The steamer itinerary, though it changed over the years in small ways, was fairly consistent from the beginning of Cook's steamer service to the end of it in the mid-twentieth century.[18] Travel by this method was not as fast as the train, given the flow of the river and all of the stops at sites along the way, but it was supposed to be leisurely and enjoyable. It allowed for visitors to take their travel a bit more slowly and to see parts of the country that travellers by train would not be able to see, all at a relatively affordable rate. Baedeker's *Egypt* informed travellers that Cook's steamers were the 'best on the Nile' and that the 'company met with on board the steamers is generally unexceptionable, though, of course,' they warned, 'it is always wise to use some little exertion to secure an agreeable and sympathetic cabin companion' if one were travelling alone.[19]

The itinerary below is taken from Cook's 1876 edition of *Tourists' Handbook for Egypt – the Nile and Desert.*[20] The book was useful as a guide for those travelling by either dahabeah or steamer, but Cook's made sure to point out, after a long list of pros (for taking a steamer) and cons (of taking a dahabeah), that steamers were the best way for most anyone to go. It is important to note here that Cook did not own dahabeahs of their own until the early 1880s, so if people decided to travel by private boat, it would have taken away business from the package-deal company. Cook's did tell travellers, however, that, 'Messrs. Cook & Son are quite prepared to book travellers either Steamer or Dahabeah, and to make the best arrangements possible to be made by either mode of transit' (Figure 3.2).[21] Cook also pointed out that the steamers, because they travelled quickly, would travel only during the day so the scenery would not be lost at night or to sleep.

The itinerary usually went as follows in Table 3.1.

These trips were three weeks of sightseeing at a blistering pace, and the steamer would arrive back in Cairo right on time. Given the fact that most tourists would need at least a week (for those based in Europe, two weeks or longer for Americans) to make the trip back home, this journey would be, at minimum, a five-week trip. If you went to the Second Cataract, it would take between seven to ten weeks. Clearly a steamer trip was meant for those with some leisure time and extra money to spend, although not as much time as needed for a dahabeah.

Into the twentieth century, Cook's continued to offer three weeks on the Nile, from Cairo to the First Cataract and back, 'First Class throughout' for £50.[22] Prices went up over the next fifty years, but these trips consistently

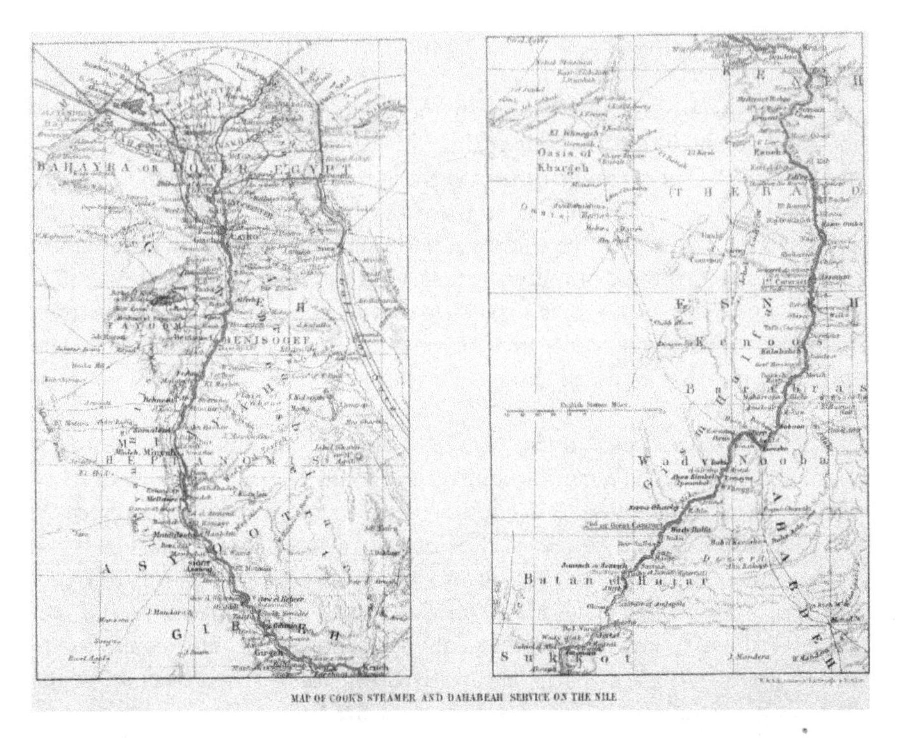

Figure 3.2 Map of Cook's steamer and dahabeah service on the Nile, 1897.

Table 3.1 Steamer itineraries

Day 1: Cairo	Leave Cairo from the docks near the Kasr el Nil bridge, at 10:00 am.[a]
Day 2: Saqqara	Reach Saqqara, about 10–12 miles from Cairo for a morning tour of the pyramids there. By noon, back on the steamer to reach Wasta for the night
Day 3: On the River	Early start in the morning so to arrive at Minieh by the evening for a visit to the Viceroy's palace which was also a sugar factory.
Day 4: Beni Hasan and River	Two hours into the day's steam reach Beni-Hassan to see the rock tombs there, but continue on to reach Manfaloot 'a town of some importance' by the end of the day.
Day 5: Asyut	Arrive at Asyut in the morning and take in the bazaar and the hills behind the town.
Day 6: On the River	On the Nile, no stopping at any points of interest

Table 3.1 Steamer itineraries (Continued)

Day 7: On the River, Keneh	Passing Abydos, to view on the return trip, the steamer reaches Keneh by evening. 'The best porous jugs and gargoulets for filtering the Nile water are made here; Keneh is also celebrated for its dates and was once noted for its dancing girls.'[b]
Day 8: Denderah	Visit to the Temple at Denderah, done by noon, so the boat may get to Luxor that evening. The boat will 'anchor opposite the Temple of Luxor'.[c] Travellers had the option to go to the temple to see it by moonlight.
Day 9: Luxor	Visit to Biban el Muluk, Valley of the Kings.[d] 'Only two or three of the most interesting need to be entered – Belzoni's Tomb, No. 17, and Bruce's, No. 11.'[e] Whether they choose to walk over the mountain or return the way they came, visitors will stop by the Ramesseum. They can also visit the Temple of Amenhotep III and the Colossi.
Day 10: Luxor	Over the river again to Medinet Habu, lunch, then visiting tombs. 'If still early, a walk through Luxor and an exploration of the Temple, mostly hidden by Arab dwellings and stables, will finish the day.'[f]
Day 11: Luxor	Temple of Karnak all day. Either eat lunch at the temple or come back to the boat.
Day 12: River, Esneh, Edfu	Leave Luxor to reach Esneh and visit its temple, then continue to arrive at Edfu by the evening.
Day 13: Edfu, Kom Ombo	Visit the Temple of Edfu, possibly, or decide to visit it on the way back. Move along to Gebel el Silsileh. 'It is hardly worth while to go ashore, since the few grottoes and chapels can just as well be seen from aboard…'[g] Two hours more of sailing and the steamer stops at Kom Ombo.
Day 14: Aswan	Reach Aswan early in the morning. 'As soon as the steamer stops, a swarm of Nubians come to offer ostrich feathers, eggs, and other articles, for sale.'[h]
Day 15: Aswan, Philae	Go through the town and bazaar of Aswan, then to the ancient quarries. Visitors could go to Philae and then the group may split here. Some may travel further up the Nile and would be taken to the Second Cataract steamer. The trip from the First to the Second Cataract would take another twelve days. Those who were going back down the Nile were taken to view the Cataract and ride a few of the rapids in a dahabeah hired for the purpose.

Table 3.1 Steamer itineraries (Continued)

Day 16: Return journey	Start the return journey, stopping at Esneh if needed and arriving at Luxor for the night.
Day 17: Return Journey	Luxor to Bellianeh
Day 18: Return Journey and Abydos	Bellianeh for excursion to Abydos, then back on board for the rest of the journey.
Day 19 and 20: Return	'Continue the voyage without stopping, except at such places as were not visited according to programme during the up journey; the steamers only stop to coal, or for provisions, according to requirements.'[i]

[a] It is important to note that all of Cook's steamer and dahabeah journeys began here.
[b] *Cook's tourists' handbook for Egypt* (1876), 18.
[c] *Ibid.*, 19.
[d] Cook built a rest house in the Valley by 1895.
[e] *Cook's tourists' handbook for Egypt* (1876), 19. Belzoni's Tomb is better known as KV 17, the tomb of Seti I. It was uncovered by Giovanni Belzoni in October of 1817 and is well known for its well-preserved decoration. Bruce's tomb is KV 11, the tomb of Ramesses III and has been open since antiquity. James Bruce entered the tomb in 1768.
[f] *Cook's tourists' handbook for Egypt* (1876), 20.
[g] *Ibid.*
[h] *Ibid.*
[i] *Ibid.*, 22.

left at 10:00 am sharp from Kasr el-Nil Bridge, every two weeks. Steamers from the First to Second Cataracts did not begin operating until January each year, lasting until March. If travellers wished to add a week to the journey, usually either at Aswan or Luxor in one of Cook's hotels, the price would increase by £15.[23] It is important to note here that the cost of this trip, for first-class accommodations, would have cost the equivalent of over six months of work for a skilled tradesperson in this same period.[24] Cook's steamers were always touted by Cook's themselves as being the best choice for those wanting a worthwhile journey for the best value. Baedeker's also recommended them 'for the immense majority of travellers ... who have not more than three or four weeks to devote to a visit to the Nile Valley and the monuments of the Pharaohs.'[25] There were cheaper Nile cruise options for those who could not afford Cook's through the other companies in business at that time, including Tewfik Nile Navigation and Henry Gaze & Son.[26] Their services were comparable to Cook's while on board, with food, excursions, and more. But the infrastructure that Cook's had built within the country in concert with the British colonial protectorate and Egyptian Government meant that Cook's, more than the others, were able to withstand war, revolution, and down seasons (if there were any). Tewfik, Gaze, and other small travel companies were not as well established as Cook's and did not have government contracts, so were out of business by the coming of the First World War in 1914.[27]

By 1895, along with Murray's, Baedeker's, and Cook's guides, there were so many travellers who had gone on steamer trips – many of them on a Cook's steamer – that there were dozens of individually published travelogues that offered tourists advice on how to travel up the Nile. These books, along with more stable political and economic foundations in Egypt, helped to add to the stream of visitors making Egypt a vacation destination. They were flocking to see antiquities, often crowding into places and leaving trash and other remnants of their visits behind. However, thanks to the influx of tourists, the fictional Egyptologist Radcliffe Emerson, husband of Amelia Peabody, expressed his exasperation to his brother in a telegram: 'Could strangle T. Cook for starting his tours. Egypt overrun with babbling idiot tourists.'[28] And he was not the only one, real or fictional, who felt this way. Amelia Edwards' disparaging talk of Cook's steamers being so 'comfortless' and 'crowded with tourists' that most travellers in fact dreaded the mass cruises, rankled Cook so much they reviewed her travelogue, poorly, in their magazine *The Excursionist.*[29]

In 1910, French naval officer and fiction author Pierre Loti wrote a book about his travels through Egypt.[30] Throughout he showed a general disdain for visitors from the Western world (of which he was one) and all tourists travelling up the Nile (of which he was one). Mostly, he railed against the hundreds of steamer tourists, 'Cooks and Cookesses' as he called them, who were 'vulgar ... barbarians', who portrayed an 'excess of grotesqueness in profanation' throughout their travels.[31] He described one group who had been lunching in the Temple of Seti I in Abydos, on the eighteenth day of their exhausting twenty-day journey, when he encountered them:

> Behold a table set for some thirty guests, and the guests themselves – of both sexes – merry and lighthearted, belong to that special type of humanity which patronises Thomas Cook & Son (Egypt Ltd.). They wear cork helmets, and the classic green spectacles; drink whisky and soda, and eat voraciously sandwiches and other viands out of greasy paper, which now litters the floor. And the women! Heavens! what scarecrows they are! ... Let us escape quickly ...[32]

It should be noted that it was because of tourists – mostly brought by Cook in this period – that travel writers had an audience for their work and that Egyptologists found generous donors, so Loti and others who complained about them were walking a tightrope.

There were a number of tourists, who visited Egypt for the first time on a steamer cruise, who went on to become influential in Egyptology. Margaret

Benson, known to most as Maggie, took her first trip to Egypt in 1894.[33] All of her trips from 1894 to 1897 and one final trip in 1900 involved a steamer journey and a long stay in Luxor to help treat her pulmonary ailments, her generalised anxiety, and likely depression.[34] She and her brother, Fred, usually booked passages on a Cook's steamer so they could enjoy the scenery on their way to Luxor. They remained in Luxor in order to rest and, later, to excavate (see Chapter 4). While most of her published letters about Egypt focus on her time in Luxor, she wrote home about their time on the boats, too.[35] They both loved the respite that cruising time gave them, and Maggie did whatever she could to make the journey seem to last as long as possible. In 1895, in fact, there were so few people on their steamer – maybe five on a boat meant for thirty – that Fred and Maggie took advantage of the time not to be forced to talk to anyone else. That year, apparently, Fred wanted to be left so unaware of his surroundings that he asked Maggie not to point out any objects of interest to him as they cruised. Then, one day, they came upon eighteen massive vultures fighting over a giant animal carcass on the shore. Maggie could not be sure what they were eating, but she wanted Fred to see it. Fighting all of her normal travelling habits, she did not tell him about the scene even though it was just feet away from them. Maggie slept and read a lot and took the time to learn Arabic and how to read hieroglyphs.[36]

Helen Mary Tirard published a detailed account of her journey on the Cook's steamer *S.S. Rameses the Great*.[37] Tirard was one of the honourary secretaries of the Egypt Exploration Fund when she took a steamer trip with her husband Charles in 1888.[38] Their boat was known as 'the largest and best of Cook's steamers on the Nile. She carries sixty-two passengers' and the boat was full for Tirard's journey.[39] Tirard's travelogue about her journey, *Sketches from a Nile Steamer: For Use of Travellers in Egypt*, was written to get travellers excited about ancient Egypt and to 'smooth their way through the confused chaos which the Egyptian temples often present' by providing plans of temples and information along the way.[40] Beginning the journey in Cairo, the steamer left on 31 January, at 10:00 am no doubt, and followed the general itinerary detailed above. The boat took two weeks to reach Philae, above the First Cataract, and then they boarded the steamer *Sethi*, which would take them to the Second Cataract and back. They returned down the Nile on the steamer as far as Asyut, where on 3 March they resolved 'to hasten back by train to Cairo to meet some friends, reluctant as we are to leave the pleasant waterway for the hot and dusty railway'.[41] The journey by train would have only taken hours, not days, and their Nile journey abruptly ended. Upon her return to England, like many women travellers at the time, Tirard continued to have an impact on Egyptology beyond her *Sketches*. She translated Adolf Erman's *Aegypten und aegyptisches Leben* into English as *Life in Ancient Egypt*.[42] She was a noted lecturer and gave a number of talks and demonstrations at the British Museum. She was also an influential member of the EEF Committee, not just with her work but also with her money, donating hundreds of pounds from 1887 until her death in 1943.[43]

Norma Lorimer, one of the most notable women authors from the Isle of Man, travelled on the same boat as the Tirards, but in 1907. Her journey, which she recorded in her travel journal addressed to her friends and family back home, was meant more to excite travellers than to educate them. In 1909, she had revised and published her journal under the title *By the Waters of Egypt.*[44] As the boat sailed, she recorded glowing memories of being on the *S.S. Rameses the Great* and how she was able to

> do nothing all day today but enjoy the Nile – do nothing but feel the spell of Egypt – do nothing but sit under a shady awning, where a cool breeze always drifts from the bows of the boat, and watch the procession of Egypt pass along the green margin of the river's banks – do nothing but watch the fierce sunlight play on the amber sands.[45]

She and her travel companion, Lorna, travelled as far south as Khartoum on the *Rameses*, then back to Aswan where they booked tickets for a train back to Cairo. Lorimer had written other travelogues like this one, such as *By the Waters of Sicily*, *By the Waters of Carthage*, and *By the Waters of Germany*, all meant to evoke a desire to travel to these places.[46] She wrote travel books and had no real investment in Egypt, though she likely influenced more than a few people to travel there themselves.

Another tourist who thoroughly enjoyed Thomas Cook's package travel was American Charles McCormick Reeve, whose father was a Civil War general and who himself was a colonel in the Spanish American War.[47] In 1890 he, his wife, and some friends from Minneapolis decided 'without much discussion' to travel to Egypt.[48] They also went to France, Italy, Syria, Turkey, Greece, and back to Italy. Reeve was a wealthy banker, manufacturer, and coin collector, so his travelogue, *How We Went and What We Saw: A Flying Trip through Egypt, Syria and the Aegean Islands* was not written to educate but to evoke and excite. To help guide his fellow travellers, he wrote:

> I have travelled with these [Thomas Cook] tickets and without them, and I say unhesitatingly that no tourist, even with the assistance of the most skilled and honest couriers (and these gentlemen are apt to be neither the one nor the other), can possibly get along so comfortably, safely, or economically in the East as when under the auspices of Thos. Cook & Son.[49]

Their steamer, the *Mohammed Ali*, left Cairo at precisely 10:00 am on Tuesday 22 January. Rather than take a dahabeah, Reeve wrote, he preferred the steamer as a dahabeah 'would not suit the average American; it may be a good way to kill time for some people, but ... seems to me a wicked waste of a portion of the alloted threescore and ten'.[50] Here he is referring to the three months of travel they carved out for a busy, but long, journey. By the end, the whole group, especially the steamer captain, was tired and irritable and they were happy to be back in Cairo, a day late.

Steamer travel was not all relaxation interspersed with tours and fun. The trips were fast and could be exhausting for passengers. Those, like Reeve, who spent the time and money to travel on steamers suffered from what some called 'pharaonic malaise', or the mental exhaustion that came from viewing too many unfamiliar sites and places. In January of 1891, as the guests on board the Cook's steamer *S.S. Rameses the Great* neared the end of their journey, they held a poetry competition to share their impressions of Egypt.[51] Many of the poetic entries were full of fond remembrances of the people they met, the things they saw, and Cook's excellent services. They remembered especially their guide Mohammed and the donkey rides to and from sites as events where 'everything ends in a donkey race'.[52]

These fond recollections were also mixed with revealing, problematic poems like this one:

If you want to know
My impressions are these
Beggars & backsheesh
Flies & fleas
Dust & donkeys
Mud & mummies
Lots of antiquities
(mostly dummies)
"Take it! Take it!"
Tease, tease, tease.
Now it's done
I hope you don't like it.
It's enough to knock down any poet
If not the worst one,
Bother, blow it!
–P.W.M-P[53]

Possibly related to the first poet, comes a second poet with their submission:

I think Egypt is very hot.
It really is a charming spot.
I like the way the Arabs squat.
Will these impressions do or not?
I am getting tired of symbols & signs
Scratched on the walls of ancient times
But the donkey riding is A. I.
Now this poem I have done. Hurray!
–B.L.M.P[54]

Surely these travellers were tired, due to Cook's exhausting, if efficient, itinerary, but the tone in these poems was still indicative of the attitudes that Western travellers had about Egypt and its inhabitants at the time. While travellers valued the historical monuments and other material remains of the ancient past, they thought themselves elevated a level or two above contemporary Egyptians because of their skin colour, religion, culture, and language. This is not to mention that a number of their Egyptian guides spoke not only Arabic, but also at least some English, French, and German, so they could communicate with tourists. It also bears mentioning that when the Tirards and Lorimer took a steamer, it was the same boat as these dubious poets, but not the same journey. Of the complainers on steamers, Tirard wrote that there would be 'grumblers', as she called them, 'who grumble even in Egypt, and therefore would grumble anywhere'.[55]

Tourist steamers were not exactly crucial spaces in the intellectual landscape of Egyptology as a science, but they were indeed central to engaging the public interest in the field by inspiring travel writers and then other travellers. Thousands of visitors to Egypt in this period took a Thomas Cook's Nile steamer cruise, and several of them wrote and published memoirs about their trips. Memoirs like Reeves', Tirard's, and Lorimer's worked to get more people to travel to Egypt and, for some, to donate money to and become involved in local museums and Egyptological societies. At the very least, these travelogues moved more of the public to learn more about Egypt and some to visit Egypt, if they were able to afford the time and money. A few of these travellers were, at the time of their arrivals, or later became, Egyptologists or active in the field in some way. Others, like Charles Wilbour and Gaston Maspero, would travel on government or postal steamers. James Breasted recognised how useful steamers and other boats were to his work; their journeys are discussed below.

Dahabeahs

Travelling up the Nile before Thomas Cook popularised steamer travel for European and American tourists in 1869, and before it was possible to do it by train in 1898, was done by private dahabeah (Figure 3.3). A typical early dahabeah was a shallow-bottomed, wooden-hulled, forty to one-hundred-foot-long vessel. The cabin, occupying one half of the boat at the stern end, would hold three to five rooms, each with one bed, whose doors led out to a small central passage. Each bed had room under it for storage of clothing and supplies. There was a saloon on the main deck which would serve as a lounge and dining room with a big table, which, for archaeologists, could double as a workspace as the room usually contained bookshelves and other communal items. The roof of the cabin would be used as an

Figure 3.3 'Empress Eugenie's Dahabeah', c. 1869.

outdoor space with seating areas and usually had a cloth for its roof.[56] The crew slept outside, near the kitchen, which looked like a small shed on the opposite end of the boat from the cabin. In the early nineteenth century, when boats were mostly wood-bottomed, it was wise to have your boat sunk, then thoroughly cleaned to rid it of vermin.[57] Travellers such as Wallis Budge, Charles Wilbour, and Thomas Huxley would have had to have this done when they travelled in the 1860s and 1870s. Wilbour noted that once he purchased his own boat, the *Seven Hathors*, in 1886 he 'put petroleum on the timbers to kill the worms' as well as 'lined all the places where rats can come in, with the tin of petroleum cans, which is very cheap here'.[58]

The boats were relatively easy for a wealthy traveller to rent as early as the 1820s.[59] Travellers would have had to pay for boat rental, including the boat and all its equipment, then separately for the captain and crew, which would have included a cook, so it was an expensive endeavour; but it was worth it if they wished to see all the sites in Egypt at their leisure. Otherwise, people were likely relegated to Cairo, as Cook's would not arrive for tourists until 1869. While Cook's guides tended to denigrate privately rented dahabeahs as troublesome and slow, by 1895 they had a few of their own dahabeahs and ran them much the same as they would a steamer voyage but on a smaller scale.[60] In one pamphlet from that year, they told prospective travellers that passengers 'engaging the Company's Dahabeahs have the special advantage of being able to receive from the Tourist steamers constant supplies of fresh fruit, vegetables, poultry, fresh beef, &c., &c'.[61] This type of advertisement informed visitors that they could reap the same convenient benefits of the steamers without having to deal with thirty-five other strangers for four weeks or longer, if they could afford it. Further, Cook's dahabeah passengers could be assured that they did not have to bargain for their boats, crews, food, or worry about anything else – always the strongest draw of a Cook's package tour. By the end of the century dahabeahs were more commonly made with an iron hull, so would not need to be sunk to get rid of bugs and rats. Also, in iron boats, the cabin space occupied a much larger portion of the deck allowing for larger or more cabins and a larger saloon or other rooms, and the kitchens and crew space were more clearly separated from the travellers' space.

While dahabeahs were the most sought after form of transportation, due to their exclusivity and luxury, most people simply could not afford the time or expense to travel this way. Baedeker's *Egypt* recommended dahabeahs only to those who fit all three of the following criteria: those 'who desire to make a closer acquaintance with the country, who have abundance of time (to Assuan and back at least 7–8 weeks), and who are indifferent to a considerable increase of expense'.[62] According to most guides, Cook's dahabeahs rented for about 130 EGP per month, Gaze had good boats for 110 EGP per month, and there were others for monthly hire, too.[63] Per person, if the cost were split evenly among a party of four or five people, one could travel in relative luxury for only a little more than the cost of a crowded steamer trip.[64] One aspect that Baedeker's pointed out to travellers was that if you hired a travel company's boat, such as Cook's or Gaze's, you would be 'relieved from all trouble in the matter of engaging a dragoman or purchasing provisions'.[65] In the event the traveller chose to engage their own dahabeah, Baedeker's provided to-the-letter advice along with a sample contract between the captain, or *rais*, and the travellers.[66] By 1908, books were informing travellers that 'placing themselves under the care of a Dragoman', or guide, would help them to 'avoid all the petty annoyances incident to direct dealings with the natives'.[67] However, books warned, 'dragomans are inclined to assume a patronizing manner towards their employers, while they generally treat their own countrymen with an air of vast superiority'.[68] This was, no doubt, an astute observation by the colonisers who did not like being treated the same way as they treated the people whose country they were in. European and American travellers were instructed to check this attitude quickly to stop the rude behaviour of these 'mere children'.[69]

When travelling by dahabeah, visitors first had to choose the boat that would fit their needs. Many of the boats were big enough for parties of four; six would become crowded. The crew numbered between eight and fifteen men and would include the *rais*, steersman, crewmen, a cook – and for regular tourists, a dragoman who acted as a mediator and a guide. Many of the dahabeahs were similar in size, so most travellers were focused on the differences in cleanliness and quality, the cost of the rental, and finding a good captain and crew. This process could take several days. Amelia Edwards told her readers that it took her forty hours to choose a boat over the course of ten days, but her maid Jenny Lane recorded that it only took a few days.[70] Travellers were advised that they should 'make it clear to the *rais* that he is not to take any other passengers or merchandise of any kind, that the whole boat shall be at the traveller's command, and that no one was to quit the boat on the pretext of visiting relatives without asking permission'.[71] Since they would not be travelling on a steamer, they did not have to worry about people they did not know. That is, unless they had to share a dahabeah because they wished to travel privately but could not afford a whole boat to themselves. Sometimes visitors would meet fellow travellers on the boat from European ports to Alexandria, and would decide to travel with them the rest of the way. Baedeker's warned against doing this, however, telling those travellers that, compared to a steamer trip,

> greater care is required in the choice of companions for the dhahabîyeh-voyage, for the close and constant intercourse in rather narrow quarters and for perhaps two months at a time is apt to produce somewhat strained relations between those who are not originally sympathetic. The 'dhahabîyeh devil', indeed, is famous in Egypt for causing those who have embarked as friends to disembark as foes.[72]

Speaking directly to those travelling 'with scientific aims', Baedeker's advised them to 'avoid travelling with those who have no particular interest in the gigantic remains of antiquity, and who are thus constantly wishing to push on hurriedly from sheer ennui'.[73]

Also part of this long process was procuring provisions, usually in the final few days before departure to ensure freshness. The most popular spot was Turnbull's shop on the Frank Street, close to the hotels most travellers had been staying in. Travellers would buy '... biscuits, rice, pickles, lemonade syrup, tea, sugar, and jam ... linen, teapots ... feather brushes ... foot baths, insect powder, and lanterns'.[74] But as travellers like Edwards, Wilbour, Andrews, and others have noted, these purchases were useful and necessary along the way. Further, many travellers who could afford this type of travel could also afford to furnish the boats lavishly. Some, as we will see below, took a piano or a library of books on the trip, for diversion or research. Each party usually carried the flag of their country of origin, and sometimes they would design a unique-to-them flag to identify their boat as they came into various ports along the river, always sure to register them in a book at Shepheard's before they left.[75]

By the time all the luggage and supplies were loaded, which usually took a day or two, the group would be ready to sail, if the wind were with them. Dahabeah season coincided with the archaeological season. Sailing in October or November meant that travellers had the best chances for cooler weather and the prevailing winds coming from the north. That time of year also usually marked the highest level of Nile inundation, which meant boats were less likely to run aground.[76]

Once on the river, the itineraries for private boats could vary as much as the travellers themselves, but there was a generally followed itinerary for most Westerners on a Nile tour. Visiting Saqqara and Memphis on the way out of Cairo may have been an anomaly as tourists might have visited them as part of their pyramid field excursions while in Cairo, or just before they left Cairo as their boat was being prepared. If the wind was not behaving and they found themselves stuck near the village of Badrashin, then they would visit Memphis and Saqqara while their boat waited. Boats frequently got stuck for hours, often becoming unstuck only when some of the crew would jump off the boat and pull it with ropes from the shore. Further complicating things, in the worst of conditions the wind could cause a boat to capsize. Usually, because of these complications of current and weather, making sightseeing stops while travelling upriver was rare, and any pauses in travel on the up journey were often strictly for gathering food, mail, or medical supplies. There were frequent and unexpected adventures – many travellers enjoyed and remarked upon seeing a crocodile or other animals, seeing cliffs pock-marked by tomb entrances from the deck of the boat, watching the sunrise and sunset, seeing friends in other boats pass along the way, watching the crew cook, wash, sing, row, or pull.[77] Once dahabeahs reached Luxor, some two to three weeks upriver from Cairo, the group might disembark for a few days in order to allow the crew to tend to the boat. Three weeks on board, in close quarters, meant the space needed cleaning and supplies. Visitors were able to relax, get mail, meet people and see the sights, but many boats went all the way to Aswan before stopping. Some went all the way to the Second Cataract. Then, the leisurely return journey would begin.

Tourists on dahabeahs would usually stop at the same places that steamer passengers would, but in the reverse order: the First Cataract, Philae, Aswan, Luxor (for the temples of Karnak and Luxor as well as the Valley of the Kings, Tombs of the Nobles, and the West Bank temples), Abydos, Asyut (for supplies), Amarna (sometimes), and possibly Dashur, Saqqara, Memphis, and Helwan. As the dahabeahs were at the mercy and leisure of the people who rented them, few itineraries were identical. In 1905, E. A. Wallis Budge, then Keeper of Egyptian and Assyrian Antiquities at the British Museum, wrote a handbook for Cook's travellers on the Nile.[78] As all itineraries did at the time, Budge wrote about the sites in the order as though one were headed upriver, to the south. Baedeker's *Egypt* and others also described the journey upriver and gave advice on what to look for on entry into cities like Luxor and Aswan as they arrived from the north, as if on a steamer. Because most people would be on steamers or on the railway, dahabeah travellers could simply read these as they travelled in the opposite direction.

Travelling with Egyptologists

Steamers were much more public spaces than dahabeahs, and much less controllable for visitors who had specific plans for their time in Egypt. As points of interest on the intellectual landscape, steamers allowed for cheap and fast travel, but they did not perform the same role in the creation of the discipline of Egyptology as I argue dahabeahs did. Dahabeahs, in fact, played a central role in development of the discipline, both while they were travelling on the river and while they were moored. While sailing on the Nile, dahabeahs acted as private spaces which allowed for private discussion of plans and ideas among people who would be working together for at least a season. They were also spaces in which prospective archaeologists – wealthy patrons and their protégés, professors and students – could receive training in methods and theory along the way. Very often, ease of travel on dahabeahs led travellers to stop in places that were off the well-worn tourist path so they could investigate sites at their leisure, drawing, recording, hunting, and collecting objects.[79] Dahabeahs were mainly spaces for seeing new sites and meeting new people along the way to a final destination.

The power of dahabeahs in these aspects make them a central institution in the history of Egyptology. Together with hotels, dahabeahs were sites of scientific discussion, network building, and professional development, floating on the Nile.

The travellers I discuss in this book, and especially in this chapter, came to Egypt first as tourists and then became Egyptologists in one way or another. Americans Charles Wilbour, Emma Andrews, Theodore Davis, and James and Frances Breasted travelled by dahabeah up and down the Nile, sightseeing, working, and collecting as they went. Travelling by dahabeah was convenient in terms of controlling where one went and how long one stayed in one spot, but most Egyptologists would not have been able to afford to travel on dahabeahs on their own. In fact, many of them depended on wealthy donors to help fund their work and travel. Wealthier tourists, like Wilbour, Andrews, and Davis, sometimes ended up buying their own dahabeahs and returning to the Nile year after year. In fact, each of them ended up becoming important patrons for Egyptology. Wilbour bought and collected hundreds of artefacts, copied inscriptions, and helped on excavations with Maspero. Andrews and Davis funded excavations, bought and collected antiquities, and Davis was credited with big discoveries. Some wealthy tourists were already patrons of universities and museums, like J. P. Morgan (Metropolitan Museum of Art) and John D. Rockefeller (University of Chicago collections and more). The Reverend Archibald Sayce, a former Oxford Don and Assyriologist, ended up buying his own boat as well. He used it to travel up and down the Nile, often following Wilbour and copying inscriptions from various sites.

There were many travellers who only made the trip up the Nile one time, but for whom it changed life forever. Amelia Edwards was one of these travellers in 1874; Helen Mary Tirard travelled only once on a steamer up the Nile in 1888.[80] As I have shown, they each published travelogues that spurred the public interest in visiting Egypt as well as support for the EEF. Spencer Compton Cavendish, the 8th Duke of Devonshire, travelled up the Nile with his family and a doctor, Ferdinand Platt, in the winter of 1907–08 in order to alleviate the symptoms of tuberculosis.[81] While Cavendish experienced some relief, he died in Cannes on his return home. Platt, on the other hand, continued a love of Egyptology and ended up translating the Book of Genesis into hieroglyphs.[82]

Although Egyptologists, in general, could not afford to buy dahabeahs themselves, they found the houseboats endlessly practical. Gaston Maspero, the director of the French-run Egyptian Department of Antiquities from 1881–86 and again from 1899–1914, used an institutional dahabeah and steamer to travel up and down the Nile on his inspection tours, and he lived in it when moored. Both Budge and Wilbour had travelled with him on this boat, as well. In a characteristic move, the famously parsimonious Flinders Petrie told James Breasted in 1894 that the cheapest way to travel in the long-term and long-distance in Egypt was by dahabeah. Petrie's first dahabeah trip was in 1883–84 on the *Philitis* with the Sheldon Amos family. He rented the boat for £10–12 and took them up in the Delta region for a few days.[83] He appreciated the freedom of travel it gave him at a relatively low cost. Despite what our guidebooks have said about pricing, for what Breasted had planned for his first season in Egypt, he may have been able to get a cheap boat as well. He and his wife Frances also had some institutional funds available, so they rented the *Olga* on their first trip in Egypt.

Amelia Edwards wrote about the *esprit du Nil* at the start of her Nile journey in 1874. As she wrote she spoke to the exclusive club of those who had travelled or aspired to travel up the Nile, because they understood certain feelings and experiences. She informed her readers that there was indeed a hierarchy, even in this extremely privileged world of Nile travel.

> The people in dahabeeyahs despise Cook's tourists; those who are bound for the Second Cataract look down with lofty compassion upon those whose ambitions extends only to the first; and travellers who engage their boat by the month hold their heads a trifle higher than those who contract for the trip.[84]

The process of travelling up the Nile was another piece of important historical knowledge that many Egyptologists did not record, unless they were women, but many tourists did. As the rest of this chapter traces the travel of some Egyptologists up the Nile, dahabeahs will be shown as spaces of table talk, truth spots, and knowledge creation and dissemination. They are important peaks on the cognitive landscape of Egyptology in Egypt, hidden in plain sight.

Charles Wilbour and the Seven Hathors

Charles Wilbour was a wealthy American journalist-turned-lawyer who was instrumental in the infamous Boss Tweed scandal in New York in the 1860s.[85] Wilbour helped a New York politician, William 'Boss' Tweed steal between $30 and $200 million from the city of New York over the course of two and a half years.[86] Wilbour was a major beneficiary of these stolen funds, becoming wealthy himself. In 1874, a few years after Tweed's indictment by a grand jury, Wilbour and his family, knowing he might be next to be prosecuted, left the US for Paris. While in Europe he was lucky enough to train with Maspero and other Egyptology scholars, so that when he arrived in Egypt in 1880, at the age of forty-seven, he was already recognised as a renowned Egyptologist despite not having published a word.[87] In December of that year he arrived in Alexandria, seasick and needing a bit of recovery from the rough Mediterranean crossing from Marseilles. That year and after, Wilbour stayed at Shepheard's in Cairo, but he did not need the powerful context of that hotel to get him connected to a patron or a client. He did, however, use Shepheard's as a base from which to buy antiquities from dealers all over town.[88] He already had professional connections and a lot of money, so most of his career in the field took place on and around steamers and dahabeahs with well-known Egyptologists. Evidence of Wilbour's influence in Egyptology is most obviously seen through his vast collection of antiquities, donated to the Brooklyn Museum in 1916 and now on display there; but his unseen work on site and through his work with scripts can be found within his letters back to his wife, Charlotte (Lottie) Beebe Wilbour, and his mother during his time in Egypt.[89]

Early on, Wilbour travelled frequently with his friend and mentor Gaston Maspero on the museum steamer. Lottie would frequently come to Egypt to sail up the Nile with Wilbour, once Wilbour was settled there. Very often, Wilbour would accompany Maspero to various sites and help in copying inscriptions. He wrote that his room on the steamer was 'not large for a room, but it is not small for a cabin, & it is comfortable & convenient'.[90] He was known to differ with Maspero on epigraphy, try to correct Mariette's work, and influence Maspero's decisions when it came to places to excavate. Through his epigraphic work with Maspero, and the time they spent working together on the museum steamer, Wilbour became so embedded in the Department of Antiquities that historian Jean Capart later remarked that, during Wilbour's time in Egypt, 'no official excavation could be made without his direct knowledge and his immediate visit to the locality'.[91]

Using his influence, Wilbour helped Maspero figure out that it was the Abd el Rassul family who had been responsible for the 'unusually fine antiquities' that had been showing up in markets in the 1870s and 1880s. Since around 1870, Maspero and other officials knew that there had to be an unknown tomb or series of tombs that someone had found on the West Bank of Luxor, and from which someone was slowly removing remarkably well-preserved antiquities. Trusting Wilbour's judgement on antiquities (and needing to continue to cultivate his patronage), Maspero deputised Wilbour in finding the culprit. In April of 1881, as Wilbour waited for Maspero to arrive in Luxor, he met with an antiquities dealer named Ibrahim Mohassib, who lived in the village of Sheikh Abd al-Qurna, on the West Bank of the river. Mohassib told Wilbour 'in a fit of confidence, that Mohammed Abd-er-Rasool, whose house is next his, found some years ago a tomb in which were £40,000 worth of antiquities; he himself saw "with these eyes" thirty-six papyri from it'.[92] Wilbour immediately telegrammed Maspero, who was given permission by the authorities to question the Abd el Rassul family in the matter. This had to be done carefully, as Mohammed was the 'right hand man' of the trusted consular agent Mustapha Agha at the time and Agha was instrumental in selling antiquities to the British Museum.[93] From there, the story is murky.

According to Wilbour, writing at the time of the event, two brothers, Mohammed and Ahmad Abd el Rassul, were taken to prison in Keneh. The head of police in Keneh was Agha's son, so they were 'not likely to suffer, and will probably be back in ten days'.[94] They were instead held for two months in prison, then released.[95] Wilbour also wrote that, simply because of the threat of torture tactics, 'M. Maspero has frightened the people, but not enough to prevent them from digging this summer, only enough to prevent them from telling him where anything is'.[96] Eventually, however, worried that his family would turn him in, Ahmad (or Mohammed, depending on the source) showed Émile Brugsch (Maspero's assistant) where the tomb lay, in the sheer cliffs behind the village, where they found what became known as the Deir al Bahri cache, containing around 40 coffins, including those of Amenhotep I, Tuthmosis I, II, and III, and many more.[97]

Budge, however, reported in 1890 that the brothers were arrested, dealt 'threats and persuasion, and many say tortures, [but] the accused denied any knowledge of the place whence the antiquities came'.[98] Thirty years later, Budge embellished even more and wrote that European authorities tortured the Abd el Rassuls for information, tying them to posts and beating them, but they would not confess. They were then 'thrown down on the ground, and the soles of their feet beaten with palm rods, and as they still refused to confess they were tied to seats, and heated iron pots were placed on their heads'.[99] In this version of the story, one of the brothers died and the other 'still bore the scars of the burns which he received from the heated pots on his forehead, face, and neck'.[100] In the end, it is unlikely that anyone was tortured.[101] Later on, Mohammed Abd el Rassul was given a job with the antiquities service and became a *rais* on the West Bank in Thebes.[102] A younger brother, Hussein, worked in the Valley of the Kings with Howard Carter.[103]

The Deir al Bahri cache was a substantial discovery, but Wilbour is rarely mentioned in the narrative of it due to the fact that he was not officially involved. He also never published anything, as would usually be expected from an Egyptologist, except for one paper in 1890 about Thutmoside waterway improvements.[104] Wilbour is well known now because his letters, now at the Brooklyn Museum, have been edited and were partially published by Jean Capart in 1936.[105] Through these resources, Wilbour's activities show how, in addition to travelling on the museum steamer, he also travelled on dahabeahs with many other fellow travellers, once boasting to his wife that he could 'see as many pleasant people in dahabeeyahs as I wish'.[106] He travelled with Archibald Sayce on Sayce's rented dahabeah, the *Timsah*, which had once belonged to Lucie Duff Gordon.[107] Together they would sail up and down the Nile, visiting sites, collecting objects, writing, and talking. Despite Wilbour's work and connections in Egyptology, he was a traveller himself, and depended on Cook's conveniences the whole time he was in Egypt. On his first arrival, he told his wife that he hired Cook's to get his luggage from the hotel in Alexandria to Shepheard's in Cairo, because Cook's could do it for one-fifth of what it would cost Wilbour to do it himself. He used Cook's mail steamers for communication, and sometimes for fresh food and fruit on board his own boat. But, like many others, he had his own opinion of Cook's tourists, whom he called 'Cookeeyeh', of which he had mixed feelings. Wilbour wrote to his wife from Luxor during his first trip up the river (invited by Cook himself to travel on Cook's steamer) that he and his travelling companions had joined with the tourist party to see some ruins on the West Bank. 'At lunch time the Cook's party is a convenience; at other times, the ruins are wide.'[108] It seems as though Wilbour enjoyed the convenience of Cook's boat and picnic lunch, but afterward he tried to steer clear of the large, loud, intrusive tourist group. All of this activity took place just before the main period covered by this book, but Wilbour's life and time on steamers needed to be included for its importance to the discipline.

Then, in November of 1886, Wilbour bought his own dahabeah for £650.[109] He wrote to his mother that he had to furnish the boat, and described that they had about 900 square feet in the cabin, and on deck they had 700 square feet covered by a canvas. He immediately added the American flag, but it took him longer to choose a name, the *Seven Hathors*.[110] He told his mother that the name 'lends itself to hieroglyphic decoration and Egyptians will call her "Seven," *Seba*. We find the Seven Hathors on most of the more perfect temples. When the Doomed Prince was born, the Seven Hathors came to make him a destiny. In the sky they are the Pleiades.'[111] Wilbour also added new mattresses, arranged shelves, and prepared it for sailing so that Lottie would be more comfortable. Once Wilbour had his own boat, his family came to Egypt with him more and more. Around New Year's Day in 1887, Wilbour remarked that it was 'a pleasure to have a moving house and ours is as big as the one Columbus first crossed in'.[112] They loved being able to float 'past unaccustomed and lovely scenes'.[113] Also on that day, their streamer, or flag, was designed and finally flew.

> The blue end of it was a cook's apron. A Hathor hieroglyphic is applied to it in white and then on the white and red streamers below is a hieroglyphic 7, thus: [see Figure 3.4] Altogether it is our American colors adapted to our *Seven Hathors* name.[114]

The design would then have been drawn in a ledger for the purpose at Shepheard's Hotel, as per traveller tradition.[115]

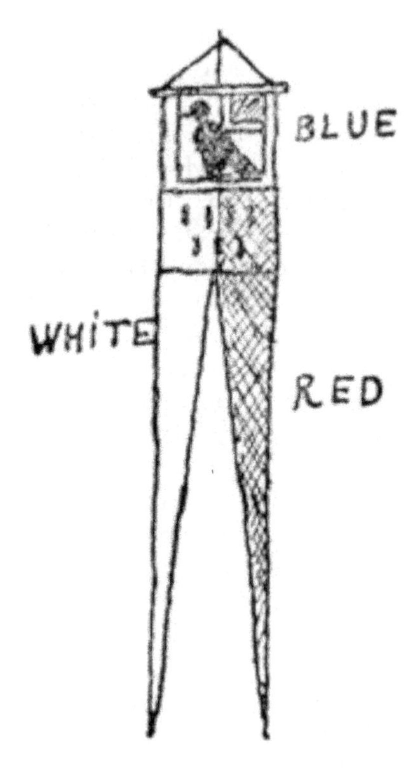

Figure 3.4 *Seven Hathors* flag, as drawn by Wilbour.

Of course, Wilbour added a reference library to rival the one on Sayce's *Timsah*. From this time on, Sayce began travelling with Wilbour, and the two would spend their days 'in copying inscriptions, and studying them in the evenings with the help of his books'.[116] His son-in-law Edwin Blashfield wrote of Wilbour and the dahabeah:

> In the center of the space was his steamer trunk, on the same was the huge folio of Lepsius and behind it on a camp-stool was the Egyptologist comparing texts. He stood discomfort wonderfully – with the mercury at one hundred Fahrenheit in February, he could spend long hours some twenty-five feet above the pavement, with his folio propped somehow between ladder and Egyptian gods in incised relief, upon the outer wall of Edfou or some other temple. Very heavy boxes of books accompanied him everywhere.[117]

Sayce recalled that Wilbour was one of the most accurate copyists he knew. For the next ten years, that was what Wilbour did.

Wilbour's impact on Egyptology was not felt in the same ways as most left their legacy – through publications or major discoveries – but instead it was through his insights into Egyptology that he shared with others while travelling on his dahabeah and the museum steamer. These boats served as floating centres of knowledge production and network building for Wilbour, Sayce, and Maspero. Sayce and Maspero silently wrote Wilbour's ideas into their scholarship and remembered him in their letters and diaries. Further, his collections, including his letters to his mother, wife, and family, were left to the Brooklyn Museum by his family in 1916. They offer an important glimpse into life on the Nile and his work in Egyptology, demonstrating not just how important dahabeahs were as nodes in the intellectual landscape, but also how important correspondence is for tracing these connections.

In the later years of Wilbour's travels, the English-language newspaper in Cairo, the *Egyptian Gazette*, noted that 'No winter visitors better understand the charms of Egypt, with the Nile trip and the opportunities for interesting research and study, than the Americans, and it must be admitted that they bring an enormous amount of money to the country.'[118] More to the point, Americans arrived late in Egypt, because it was not until after the American Civil War ended in 1865 that wealthy Americans could feasibly travel and collect artefacts further afield than Europe.[119]

James Henry Breasted

Around the time that Wilbour was ending his tenure in Egypt (he died in 1896), American James Breasted was just beginning his. After earning his PhD at the University of Berlin under Adolf Erman in 1894, but before leaving Berlin, Breasted had taken his first step in building his professional network so that he would indeed have a successful first trip to Egypt.[120] He sent copies of his dissertation, which analysed ancient Egyptian hymns to the Sun under Amenhotep IV (Akhenaten), to a number of archaeologists and Egyptologists across Europe and the UK.[121] He received favourable responses from many of them. The most exciting response Breasted received was from Flinders Petrie. By this point, Petrie had been in the field for ten years. He had established himself as an infamously frugal excavator, and he was undoubtedly the leader in excavation practices in Egypt. He had a ready group of students who he and, later, Margaret Murray, trained in the classroom at UCL and from whom he had his choice of field assistants, who became known as Petrie's Pups.[122] During the season in question, Petrie was also beginning to recruit and train the group of Egyptian workmen who he would hire for nearly every season for the next forty years: the Quftis. Aside from acknowledging Breasted's dissertation in his response, the Professor also offered, 'some kindly advice, the promise of some things for our Museum and above all an invitation to come & spend some time with him at his excavations of Coptos!!!!'.[123] Upon his arrival, Breasted found a ready mentor and Petrie had a willing new Pup.

In his letter, not surprisingly, Petrie also advised the newlyweds on the cheapest way to travel in Egypt: by boat on the Nile.[124] In accordance with Petrie's advice, and through the help of an Egyptian acquaintance, the Breasteds ordered a dahabeah, the *Olga*, on which they would sail down the river from Aswan to Cairo. The *Olga* may have been a Cook's dahabeah, and if it was, it likely cost around 130LE per month for the Breasteds.[125] Petrie's dahabeah ten years earlier had only cost around £10, so it is possible Petrie was able to arrange a better rate for the Breasteds.[126] It is likely they only spent a month on the dahabeah, as well, which may have been cheaper than renting it for a whole season. Again, this sounds expensive, but for the job Breasted needed to do for the university in collecting antiquities and visiting sites, it was the cheapest conveyance. Their dahabeah during that first season operated for Breasted as a place not only to begin his professionalisation process by building his network, starting with Petrie, but also as a space where he trained himself in field methods and prepared himself for his toughest job interview yet.

The Breasteds reached Asyut by train from Cairo, sailed up the Nile to Aswan, then they sailed down the river, with the current. They visited a number of sites along the way, including Luxor twice (going up and down) and Elephantine Island. The highlight of the trip for Breasted was getting to the Petrie camp in late December. It was such an important event for Breasted, he later recalled, that upon reaching Naqada, just below Luxor on the river, he 'jumped ashore and without waiting for a donkey to ride, hurried off on foot to find Petrie. His eagerness and the warm welcome he received made him oblivious to the long, tiring walk.'[127] Breasted found Petrie on site, dressed in a style that was, 'not merely careless but deliberately slovenly and dirty. He was thoroughly unkempt, clad in ragged, dirty shirt and trousers, worn-out sandals and no socks.'[128] Despite his appearance, Petrie was ever the professional excavator, and trained Breasted in his meticulous methods, as he would have done with one of his own Pups. Petrie and his assistant James Quibell (one of the earliest Petrie Pups) were already ensconced in the site, excavating a large pre-dynastic cemetery.[129] Because of the massive number of burials – along with the group of men from Quft, they excavated over 2,200 graves – Petrie had quickly developed an organised system of uncovering and safely cleaning each grave, recording the finds, and collecting the objects.

He recorded this system in the field report for the season in order to 'give sufficient confidence in the general accuracy of the results noted'.[130] He explained that he used a 'compound gang' of pairs of Egyptian men and boys, the Quftis, led by his *rais* (foreman) Ali Suefi.[131] He described the process:

> First a pair of boys were set to try for a grave, and if the ground was soft they were to clear around up to the edges of the filling, but not to go more than a couple of feet down. At that point they were turned out to try for another, and an inferior man and boy came in to clear the earth until they touched pottery or bones in more than one place. They then turned out to follow where the boys were working, and the pair of superior men came in to dig, or to scrape out with potsherds, the earth between the jars. While they were at work Ali was in the hole with them, finishing the scraping out with a potsherd or with his hands, his orders being to remove every scrap of loose earth that he could without shifting or disturbing any objects. When he had a favourable place his clearing was a triumph; every jar would be left standing, still bedded to the side of the grave, while all the earth was raked out between one jar and another; the skeleton would be left with every bone in its articulations, lying as if just placed on the ground, the cage of ribs emptied, and the only supports being little lumps of earth left at the joints. The flint knives or other valuables would be each covered with a potsherd, to keep it from being shifted and a pebble laid on that, to denote that it marked an object.[132]

Petrie would then come to the grave to record the locations of the objects as well as finish the removal of the grave goods and skeleton. His excavation and recording techniques were routinely detailed and usually focused on small objects such as potsherds and necklace beads. From these objects, Petrie was able to devise a dating system for ancient Egyptian history, thus giving him the nickname 'Abu Bagousheh – Father of Pots'.[133]

Breasted spent almost two weeks with Petrie on the site and 'absorbed every detail of the technique of excavation, its supervision and cost', as a good Pup was expected to do.[134] Not only did he witness Petrie's new cemetery excavation technique in detail, but he also learned that the previous year, 'Petrie had paid "just five shillings a week for provisions for himself and his assistant"'.[135] Before leaving Naqada, Petrie had suggested to Breasted that they should collaborate on an excavation site for a season, and Breasted was happy to consider it. As an American, Breasted had the potential to bring a lot of private money with him, amounts which the British were not able to secure.[136] However, while Breasted believed that excavation was 'eminently worth-while', it was only of 'secondary importance' to him.[137] Instead, over the course of his Nile journey, while he read and prepared himself for his work on board the *Olga*, he 'foresaw that his own most important work in Egypt would be the reconstruction of her ancient past rather than the recovery of the material remains of her civilization'.[138] This reconstruction project soon became his life-long goal, institutionalised in The Epigraphic Survey at the University of Chicago, whose continuing mission is to record all surviving inscriptions on temple and tomb walls in Egypt and publish them before they perish with time.[139] Petrie not only taught Breasted how to excavate, but also allowed Breasted to realise that excavation was not his passion or purpose. This part of Breasted's training did, in fact, take place at an archaeological site. However, because he was able to travel on the dahabeah, he was able to arrive there and do the work that would set him up for the next decade as an author and professor in Chicago.

Breasted depended on dahabeahs even more after he had established himself as an Egyptologist in the US. He continued to use them for the rest of his career as spaces in which to do science, to train his epigraphers, and to communicate his results to donors and other scholars. In 1905, he returned to Egypt after a ten-year hiatus with Frances and their son Charles, who was eight. At this point, Breasted was the Director of the Haskell Oriental Museum, held the chair of Egyptology and Oriental History as a full professor at the University of Chicago, and had just been awarded a generous grant by the John D. Rockefeller Foundation to begin the Epigraphic Survey he had envisioned on his first dahabeah journey. As detailed in the previous chapter, while Breasted's plan that season was to ask for permission to 'photograph or copy all the inscribed ancient monuments of Upper Egypt', Maspero instead convinced him instead to 'select a special district in Nubia and to work it out before asking for a second one'.[140] He also told Breasted not to clean or excavate any of the sites.

To prepare for the journey, Breasted collected photographic supplies, lamps, books, dry goods provisions, medical supplies, and other equipment.[141] Much of this equipment was brought from Berlin, as he could not get the appropriate photographic technology in Cairo. They also visited with Emma Andrews and Theodore Davis on their dahabeah, the *Bedauin* before they left.[142] As he had done a decade earlier, he rented a dahabeah near the First Cataract so they could sail into Nubia.[143] The group, comprised of the Breasteds and two other crew members – engineer Victor Persons and photographer Friedrich Koch – likely took a train to Aswan where they would have met their boat. By the end of December, they were aboard the dahabeah *Abu Simbel*, and sailing further upriver to the Second Cataract and further to their first site at the Wadi Halfa.[144] The *Abu Simbel* was an eighty-foot-long boat with one large cabin for Breasted and his wife, and four smaller cabins, one for Charles, one each for Persons and Koch, and one reserved as a dark room for processing photographs. According to Breasted, the *Abu Simbel* was 'a good boat, but we are heavily loaded, having over a hundred boxes, trunks, bales, &c., packed away on board'.[145] Breasted enjoyed his time on dahabeahs so much that he wrote to his mother on 30 December 1905 that 'I was never better in my life, and the trip is evidently doing me a world of good'.[146]

They reached Wadi Halfa on 7 January 1906 and started to work the following day. The process by which they worked to copy the temple inscriptions is now known as the Chicago House Method, and was then as it is now, one of the most accurate ways to copy crumbling inscriptions. The wall surface is carefully photographed with a large-format camera, then printed the same day, hence the need for the dark room on board. The large-format prints are compared to the temple wall surface, and corrections are made in red pencil right on the print at the site. The most accurate drawings come from these prints.[147] After about a week at Wadi Halfa, the crew made its way back to the site of Abu Simbel, where they worked for five more weeks.

Breasted encountered a number of challenges in terms of getting good lighting and angles for the photographs. At one point, Breasted had to tie himself to the top of the dahabeah's masthead to get a parallel shot of one wall, and at another point he had to lower himself head-first into a trench dug for the purpose. At the end of this particular season, Breasted and his crew had surveyed eighteen major structures, about which he remarked that '*all* the pre-Ptolemaic monuments of Lower Nubia are included and completed'.[148] Much like he would have done in a classroom or his office at the University of Chicago, Breasted was able to train his crew, plan the work, and write up results in the time they were sailing to and from the site. Everyone was able to get their equipment set up on board and work quickly and efficiently. The dahabeah functioned as both a work space and a home space, and, while during this trip Breasted did not entertain dignitaries on board, some of his later journeys in the late 1920s and early 1930s were designed to do just that.

Breasted had such a successful season on the dahabeah that by April of 1907 he had proposed a 'floating headquarters and working laboratory' on the Nile.[149] He submitted his proposal for a three-storey, eighty-foot-long, twenty-two-foot-wide, steel-hulled steamer that would cost around $14,000 to build in Cairo to Frederick T. Gates, the financial adviser for the Rockefellers.[150] Breasted had had a lot of success with the Rockefellers, and would continue to do so for decades, but this particular idea did not float with them. Seven years later, he proposed this same structure – with a few additions – to a local Chicago philanthropist, Norman Wait Harris. Breasted argued that the boat would serve a number of purposes, including having 'people of means' come to visit to see the work Breasted was doing, as well as hosting lectures, dinners, teas, and other social and scientific gatherings.[151] Breasted readily recognised and emphasised the fact that this boat, led by himself and with a large library on board, would quickly 'become a recognized institution' and 'a great archaeological institute ... a centre to which the scientific men of the whole world would habitually turn for authoritative research' in Luxor and allow for the Epigraphic Survey to continue its work in the region.[152] But Harris also denied this request, and although the boat itself was never to be, Breasted was convinced he should continue to use dahabeahs as floating labs for his research.

For the next several years, Breasted depended on dahabeahs both as central work sites and as transport from one work site to another for the Epigraphic Survey. In 1922, he and Charles, who by that time was a University of Chicago graduate and sometimes-reluctant assistant, travelled by dahabeah which reached Luxor just in time for the uncovering of the tomb of Tutankhamun (see Chapter 4). In 1929, Breasted and Charles boarded a chartered steamer for a very different reason – to vacation with John D. Rockefeller, Jr.[153] All they had planned to do was see the sights and enjoy themselves, as Breasted remarked, and he looked forward to relaxing. But he was also on display by his biggest patron, so he knew he had to use that particular trip to maintain this important relationship. Breasted used these boats in a number of ways and his time in Egypt is a clear example of boats as institutional spaces for the practice of Egyptology.

Emma Andrews and Theodore Davis

In terms of important donors to the practice of Egyptology, Emma Andrews and Theodore Davis did not only give money, they also gave time and resources both while they were in Egypt as well as when they returned to the United States. Their collections are housed in various places, but most of all in the Metropolitan Museum of Art in New York.[154] They were regular guests at Shepheard's Hotel and often met their European crew there before heading to the field (see Chapter 2). They excavated primarily in the Valley of the Kings, thanks to being the sole concession holders for over a decade (see Chapter 4). In order to get from Cairo to Luxor, and back, they first rented, then eventually had a dahabeah custom built.

Andrews' diary is an invaluable resource for their time in Egypt, the work they did, and the impact they had on Egyptology. Her diaries began in 1889 on their first trip up the Nile, but the pair had been in Egypt in 1887. Andrews even commented in the beginning of her diary that she 'felt it rather strange and interesting that we should have familiar recollections of Egypt'.[155] They had planned, originally, on chartering a Cook's boat, but they 'were not pleased' with the 'little steamer' offered by Cook's, the *Sethi*.[156] According to Cook, the boat had the capacity for at least twenty-six passengers, and was frequently booked for private trips to Luxor, Aswan, and beyond.[157] It is possible the couple may not have liked the fact that they were not able to control the ship's itinerary or the presence of other guests if they were unable to book it for a private trip. By Christmas of 1889, they had decided to rent a private dahabeah instead. Andrews wrote of their choice: 'Theodore decided upon the "Nubia" belonging to Prince Achmed, an iron boat, with an excellent reputation for speed, and with pretty fittings, and very good beds. Her model and fittings were so superior to those of any others that we saw, that in spite of her high price, Theodore decided upon her.'[158] Andrews did not go into detail about the price or all of the preparation they did to get ready for sailing, but she wrote that their maid, Georgette, 'had packed all our belongings, and we were more or less ready – our own trunks, and innumerable big hampers, mysterious rolls of wraps and carpets, were all piled in the hall by our doors' at Shepheard's.[159] They also added to the boat's furniture a piano, a small library of their books, and the American flag to the mast, as was custom.[160] The group set sail on schedule, and Andrews described the end of their first day on board, 28 December: 'A lovely sunset finished the bright day, so full of novelty and pleasure – and the young moon about 5 days old made the night, when it had descended, brilliant.'[161]

They likely followed a typical itinerary going up the Nile, as quickly as others would, and the *Nubia* approached Luxor two weeks after leaving Cairo. They stopped in Luxor for one night only, on their way up the Nile. They continued up the Nile, stopping at the First Cataract, then making their way back down the Nile, slowly, stopping in Luxor again, but for a longer period. While in Luxor for the first time, Andrews remarked:

> It is very stirring and exciting to think of the constant digging and searching among the tombs across the river. Every Arab in Goornah I am told spends his night in this way. It is against all law – the government allows no digging or excavations – but they seem powerless to prevent it. It would be difficult to put such extensive territory under sufficient guard – and they have not the money to undertake thorough and intelligent excavations themselves. I feel in a rage about it when I think of it. In the meantime hundreds of hands are surreptitiously at work at it – valuable things are destroyed and injured by hasty and forbidden search ... It is soon known among this thieving fraternity, that a tourist is anxious to buy good things, and willing to pay for them.

Despite her rage, Andrews and Davis were very willing to pay well for these 'good things'. During their many years in Egypt, the pair did what many Western archaeologists did: they collected antiquities. Many times, they bought their most prized pieces from dealers. Early in their first year on the Nile, they were introduced to 'an old Mohammed Mohassib ... the most prominent dealer in antiquities in Luxor' by their *rais*, Mohammed Salah.[162] Mohassib had started his career as Lady Duff Gordon's donkey-boy in the 1860s and soon became a well-respected dealer in Luxor, opening his shop around 1880. He was well known among Egyptologists and tourists, both of whom depended on him to enlarge their personal and institutional collections. While he often dealt in stolen or illicit items, he was known to save 'his finest items for his richest customers and held his sales with great discretion'.[163] Mohassib would also give gifts to wealthy shoppers in order to remain on good terms.[164] He had a number of well-known customers, like E. A. Wallis Budge, Wilbour, Maspero, Petrie, Breasted, J. P. Morgan, and Lord Carnarvon. Davis and Andrews bought quite a few pieces from him throughout the years, such as jewellery, papyri, and statues, and sometimes pieces taken from their own sites. But the pair respected him so much that they frequently had him to tea on their dahabeah when they were moored in Luxor.[165]

In 1896, Andrews and Davis commissioned their own dahabeah to be built in Cairo, which they named the *Bedauin*. When they arrived in Cairo that December, they found the boat construction was a month behind schedule. They urged the workers to hurry, but there were 'so many vexatious delays' that Andrews got tired of Cairo.[166] She went on, 'if not [tired] of Cairo, certainly of the life of hotels – and the dreary people'.[167] But once the boat was finished in early January 1897, the furniture Andrews had bought in the US and shipped to Cairo had arrived. Nothing fit quite well enough, she said: 'Chairs too large for the rooms – everything a little too big, or a little too small – the sideboard entirely forgotten.'[168] So she had to go into Cairo to finish up the shopping. On 11 January 1897, they took possession of the boat and left Cairo as quickly as they could. The *Bedauin* was eighty feet long, had a salon with a grand piano for recitals, the dining room had a crystal chandelier, and, of course, it had a well-stocked library.[169] This boat became a fixture in the lives of Andrews, Davis, their families, and archaeologists for the next sixteen years allowing them to fully immerse themselves in the dahabeah dining society. Both Davis and Andrews flourished in this new life. Andrews wrote that year: 'I can never weary of this – night or day. Each hour has its own charm, and I do not want to leave.'[170]

By 1900, because of their wealth and their dahabeah, Davis and Andrews had met and developed relationships with a number of archaeologists, such as Briton Howard Carter, who charmed Andrews and impressed both of them with guided private visits to newly discovered tombs to see undisturbed remains and objects.[171] Carter and Davis connected almost immediately, as both men were 'consistently honest and direct, if a bit tactless. … [they] were accustomed to talking plainly with each other'.[172] Carter was accustomed, as most Egyptologists were in those days, to wooing wealthy patrons for excavation support; he used the hotels and the dahabeahs to do so. At the same time, British Egyptologist Percy Newberry was joining them on the *Bedauin* for visits, tea, and talk. Andrews funded some of Newberry's excavations on her own, later. Later on, they also met Arthur Weigall, the inspector for Upper Egypt after Carter had left the post, and Weigall quickly moved into the role in Davis' and Andrews' world that Carter had vacated. From 1900 to their last season in 1913, the pair were instrumental in archaeology in Egypt and many of the most important finds in the Valley during this period can be attributed to them and their patronage. Much like other travellers in this chapter, it was their first journey upriver by dahabeah in 1889–90 that shifted Andrews' and Davis' world in two main ways: first, they began their love affair with Egypt that would last more than twenty years; second, they both began their archaeological careers. Andrews became a collector, traveller, and diarist, and her excavation records are usually the only field notes we have for their archaeological (or many other) activities. Davis also became a collector, and, due to their joint interest in the archaeological record in Upper Egypt, the pair quickly became known to the Antiquities Department as generous and knowledgeable patrons. The pair, their work, and their boat will be central characters in the next chapter.

Conclusions: The dahabeah dining society

No matter their institutional affiliation (or not), seeing Egypt from the deck of a dahabeah was the first way many of the archaeologists in this book experienced and became enamoured with Egypt. Jason Thompson points out that these vessels 'remained an important part of Egyptological life in Egypt throughout the nineteenth century'.[173] In this part of the voyage, I have discussed just a few of the many trips that archaeologists had taken up and down the river. I focused mainly on first journeys because they tended to be the most transformative, inspiring art, literature, travelogues, and decades of Egyptological work. These boats were homes away from home, as well as spaces that created Egyptologists or archaeologists by the end of a journey embarked upon by a tourist.

To say that these boats were places where the public and private would meet would be inaccurate. In fact, dahabeahs were much more exclusive than hotels, and could be put in the same liminal category as dig houses.[174] Boats and houses were places where only those chosen few who were invited could participate. Wealthy patrons regularly fed and housed their Egyptologists on dahabeahs; royalty and other tourists of distinction were wined and dined on dahabeahs as well. When he was the Director of Antiquities, Maspero was a fixture on others' boats, hosted by visitors while he sailed on the museum steamer. Regular tourists, that is the Cooks and Cookesses, on the other hand, would not have been invited, so the amount of work and level of discussion about Egyptology, artefacts, and more would have been different than the work done at hotels in the proximity of the vacationing public. Breasted's work on dahabeahs showed that they were also training sites for new crew members and new visitors to Egypt as a classroom or museum work space may be. By examining many of these activities, including itineraries, we have seen who was participating, who was allowed to participate, and who was recorded as an important contributor, we have travelled along with these people, experiencing all that Egypt's river life had to offer.

When boats were moored at sites, there was a lot more activity happening on them than when they were sailing. During Breasted's journeys, the dahabeah was a central place for knowledge creation through excavation, photographs, and the Epigraphic Survey when it was moored at Naqada, Aswan, and Abu Simbel; it was also a crucial vehicle for travel between sites. Wilbour used his time on the museum steamer, his dahabeah, and others' boats as an opportunity to learn Egyptology, make connections, and establish his own authority in a field in which he otherwise hardly made a mark when he was alive. And yet, he is still recognised as a central figure in early Egyptology because of the work he did and the collection he amassed while travelling up and down the Nile. Davis and Andrews were famous in Egyptology for their patronage and their hospitality on their own boat, the *Bedauin*. Metropolitan Museum of Art curator Albert Lythgoe noted at the beginning of the typed version of Andrews' diaries that the 'charming description which she gives of their river-life' was 'familiar to many of us who enjoyed their hospitality'.[175] Without going through dozens, possibly hundreds, of other archives, it is impossible to know the impact that Andrews and Davis had upon all of their visitors, but we can plainly see the impact that their river journeys had on Egyptology. Their collections at the MMA, along with Wilbour's at the Brooklyn Museum just eleven miles away, made New York a centre of Egyptological knowledge.

As travellers arrived in Luxor for a long stay, hotels again became central nodes in the intellectual landscape of the Western study of Egyptology. Dahabeahs and dig houses will also play a role in Luxor, as collections were built, knowledge was shared (among the privileged), and Egyptians were largely excluded from any of this activity by virtue of the places and the people involved.

Notes

1 Edwards, *A thousand miles up the Nile*, 58–61.
2 See Brenda Moon, *More usefully employed: Amelia B. Edwards, writer, traveller and campaigner for ancient Egypt* (London: Egypt Exploration Society, 2006).
3 Riggs, *Unwrapping ancient Egypt*, 2. John Wilson also included an entire chapter in *Signs & wonders* called 'The houseboat on the Nile (1880–1908)', which detailed the activities of Sayce, Wilbour, Davis, and Breasted on their dahabeahs (99–123).
4 This will be covered more below, in Chapter 4.
5 Baedeker's *Egypt* (1898), xxi.

6 'Cook's Cheap Express Service: Between Assiout and Assouan, Commencing December 4th 1888', Cook Archives, Black Box 1.
7 *Ibid.*
8 See the conversion Table 0.1. These amounts are equivalent to around £1,739 today, or $2,404.90 for the eleven-day trip; £2,174 today, or $3,006.30 for the fifteen-day trip; servants cost £869.60 or $1,202.44 today for the eleven-day trip and £1,304 or $1,804.66 for the fifteen-day trip.
9 'Cook's Cheap Express Service: Between Assiout and Assouan, Commencing December 4th 1888', Cook Archives, Black Box 1. More on this in Chapter 4.
10 See the conversion Table 0.1. These amounts are equivalent to between around £1,391 or $1,925 and £2,174 or $3,006.30 today.
11 Baedeker's *Egypt* (1898), xxxi.
12 'Independent Tour itinerary for a Mrs. Luhr, beginning 19 September 1911', Guardbook 13, Cook Archives.
13 Baedeker's *Egypt* (1898), 177.
14 *Ibid.*
15 *Ibid.*
16 W. E. Kingsford to Emily Paterson, 12 November 1894, EES.XId14.
17 *Pamphlet: Thos Cook & Son in Egypt 1869–1926: History of the company in Egypt.* Cook Archives, Black Box 1.
18 I will talk a bit in this chapter about Thomas Cook's steamers and other boats. For an excellent, in-depth history of Cook on the Nile, see Humphreys, *On the Nile.*
19 Baedeker's *Egypt* (1898), 28.
20 *Cook's tourists' handbook for Egypt, the Nile, and the Desert* (London: Thomas Cook & Son, 1876), 16–23.
21 *Cook's tourists' handbook for Egypt* (1876), 16.
22 *Cook's Excursionist and Tourist Advertiser* (16 September 1895): 17, Cook Archives. See the conversion Table 0.1. In 1900, £50 is equivalent to £4,142 or $5,732 today.
23 *Cook's Excursionist and Tourist Advertiser* (16 September 1895): 17, Cook Archives. See the conversion Table 0.1. In 1900, £15 was equivalent to £1,243 or $1,720 today.
24 See the Currency Converter at the National Archives UK, which calculates value and purchasing power: £50 would be the equivalent of 151 days of work for a skilled tradesperson (www.nationalarchives.gov.uk/currency-converter/ accessed 22 April 2021).
25 Baedeker's *Egypt* (1892), xiii.
26 *Ibid.*, xv
27 Brendon, *Thomas Cook.*
28 Elizabeth Peters and Kristen Whitbread, eds, *Amelia Peabody's Egypt: A compendium* (New York: William Morrow, 2003), 14.
29 Edwards, *A thousand miles up the Nile*, 36, 100. *The Excursionist* in 1877 reviewed Edwards' book unfavourably (Humphreys, *On the Nile*, 60–1).

30 Loti, *Egypt (La Mort de Philae).*
31 *Ibid.*, 141; 136–7.
32 *Ibid.*, 136–7.
33 Benson, ed., *Life and letters.*
34 Kathleen Sheppard, '"Constant Companions" and "Intimate Friends": The Lives and Careers of Maggie Benson and Nettie Gourlay,' *Lady Science* (57): 6 June 2019 (www.ladyscience.com/constant-companions-and-intimate-friends/no57 accessed 13 October 2020).
35 Benson, ed., *Life and letters.*
36 *Ibid.*
37 H. M. and N. Tirard, *Sketches from a Nile steamer* (London: Kegan Paul, Trench, Trübner & Co., Ltd., 1891).
38 Amara Thornton, *Archaeologists in print: Publishing for the people* (London: UCL Press, 2018), 51–2.
39 Tirard, *Sketches*, 1.
40 *Ibid.*, vii.
41 *Ibid.*, 211.
42 Adolf Erman, *Aegypten und aegyptisches Leben* (Tübingen: H. Laupp'sche Buchhandlung (1885); H. M. Tirard, transl. *Life in Ancient Egypt* by Adolf Erman (New York: Macmillan & Co., 1894).
43 Thornton, *Archaeologists in print*, 51; Bierbrier, *Who was who in Egyptology*, 542.
44 Norma Lorimer, *By the waters of Egypt* (London: Methuen & Co, 1909).
45 *Ibid.*, 169.
46 Norma Lorimer, *By the waters of Sicily* (New York: James Pott & Co., 1901); Norma Lorimer, *By the waters of Carthage* (New York: James Pott, & Co., 1906); Norma Lorimer, *By the waters of Germany* (London: S. Paul, 1914).
47 Charles McCormick Reeve, *How we went and what we saw: A flying trip through Egypt, Syria, and the Aegean Islands* (New York: G. P. Putnam's Sons, 1891).
48 *Ibid.*, 1.
49 *Ibid.*, 3.
50 *Ibid.*, 72.
51 Cook Archives, Guardbook 13.
52 *Ibid.*
53 *Ibid.*
54 *Ibid.*
55 Tirard, *Sketches*, 1–2.
56 Humphreys, *On the Nile*, 36.
57 *Ibid.*, 34
58 Jean Capart, ed., *Travels in Egypt (December 1880 to May 1891): Letters of Charles Edwin Wilbour* (Brooklyn: Brooklyn Institute of Arts and Sciences, 1936), 411.
59 Thompson, *Wonderful things*, 150.

60 'Dahabeah Arrangements,' *Cook's Excursionist and Tourist Advertiser* (16 September 1895): 7.
61 *Ibid.*
62 Baedeker's *Egypt* (1892), xiii.
63 See the currency conversion Table 0.1. In 1895, Cook's would have cost the equivalent of £11,587 or $20,495 today; Gaze's would have cost the equivalent of £9,804 or $17,342.60 today.
64 See the currency conversion Table 0.1. The cost would have been around £2,317 or $4,100 per person per month for five people on a Cook's dahabeah; for Gaze's, it would have cost around £1,960.80 or $3,469.
65 Baedeker's *Egypt* (1892), xix.
66 *Ibid.*, xx.
67 Baedeker's *Egypt* (1908), xxiv.
68 *Ibid.*, xxv.
69 *Ibid.*, xxiv. And yet, Baedeker's did not tell the traveller what to do when they themselves acted that way.
70 Edwards, *A thousand miles up the Nile*, 11–13; J. Lane MSS 1, 6 December 1873, Journal 1, Jenny Lane Collection, Griffith Institute of Egyptology, Oxford University.
71 Humphreys, *On the Nile*, 33.
72 Baedeker's *Egypt* (1898), 28.
73 *Ibid.*
74 Humphreys, *On the Nile*, 35.
75 Gregory, 'Scripting Egypt', 134.
76 Humphreys, *On the Nile*, 38.
77 Marianne Brocklehurst, *Miss Brocklehurst on the Nile: Diary of a Victorian traveller in Egypt* (Cheshire: Millrace, 2004).
78 E. A. Wallis Budge, *Cook's Handbook for Egypt and the Sudan* (London: Thos. Cook & Son, Ludgate Circus, 1905).
79 Edwards, *A thousand miles up the Nile*; Brocklehurst, *Miss Brocklehurst on the Nile.*
80 Tirard, *Sketches.*
81 Wilkinson, *Aristocrats and archaeologists.*
82 *Ibid.*, 129.
83 See the conversion Table 0.1. This would be the equivalent of around £701 or $1,092 today. Petrie MSS 1.3.1–50, 22 November 1883, Petrie Journal 1883–1884, Griffith Institute of Egyptology, Oxford University.
84 Edwards, *A thousand miles up the Nile*, 60.
85 Susan Allen, 'Tycoons on the Nile: How American millionaires brought Egypt to America', in *Lost and now found: Explorers, diplomats and artists in Egypt and the Near East*, eds Neil Cooke and Vanessa Daubney (Oxford: ASTENE, 2017), 71.
86 That is equivalent to between $365 million and $2.4 billion today.
87 John M. Adams, 'A Bad Dream of New York: The Rise, Fall, and Redemption of Charles E. Wilbour', n.d.

88 Charles Wilbour to Lottie Wilbour, Sunday, 30 October 1881, Folder Letters 4.1.002 October 1881–April 1882, Wilbour Papers 4.1, Brooklyn Museum.
89 Capart, ed. *Travels in Egypt.*
90 Charles Wilbour to Lottie Wilbour, 15 December 1881, Folder Letters 4.1.002 October 1881–April 1882, Wilbour Papers 4.1, Brooklyn Museum.
91 Capart, *Travels in Egypt*, vii.
92 Charles Wilbour to Lottie Wilbour, 4 April 1881, Folder: Letters 4.1.001 December 1880–April 1881, Wilbour Papers 4.1. Brooklyn Museum. See conversion on Table 0.1. This would be the equivalent of about £2,805,600 or $4,238,000 today.
93 Charles Wilbour to Lottie Wilbour, 8 April 1881, Folder: Letters 4.1.001 December 1880–April 1881, Wilbour Papers 4.1. Brooklyn Museum.
94 *Ibid.*
95 Budge, *The Nile*, 582.
96 Charles Wilbour to Lottie Wilbour, 8 April 1881, Folder: Letters 4.1.001 December 1880–April 1881, Wilbour Papers 4.1. Brooklyn Museum.
97 Thompson, *Wonderful things*, *Vol.* 2, 8.
98 Budge, *The Nile*, 581–3.
99 Budge, *By Nile and Tigris*, 113–14.
100 *Ibid.*, 114.
101 Christina Riggs recounts this story in *Unwrapping Ancient Egypt*, and argues that they were tortured once in prison and there were no Europeans around to see it (61–3).
102 Bierbrier, *Who was who in Egyptology*, 2.
103 *Ibid.*
104 Adams, 'A Bad Dream of New York'.
105 Capart, *Travels in Egypt.*
106 Charles Wilbour to Lottie Wilbour, 1 March 1881, Folder: Letters 4.1.001 December 1880–April 1881, Wilbour Papers 4.1. Brooklyn Museum.
107 Thompson, *Wonderful things*, *Vol.* 2, 61.
108 Charles Wilbour to Lottie Wilbour, Monday 24 January 1881 from Luxor Hotel, Letters 4.1.001 December 1880–April 1881, Wilbour Papers 4.1 1880–81, Brooklyn Museum.
109 It is unclear if this is Great British pounds or Egyptian pounds. But it is likely to be GBP, so it would be the equivalent of around £45,600 today, or $71,000. See the conversion Table 0.1.
110 Capart, *Travels in Egypt*, 29 November 1886 to Lottie, 410–11.
111 Capart, *Travels in Egypt*, 13 December 1886 to Mother, 413–14.
112 Capart, *Travels in Egypt*, 1 January 1887, 417.
113 Capart, *Travels in Egypt*, 3 January 1887, 417.
114 Capart, *Travels in Egypt*, 4 January 1887, 417–18.
115 Gregory, 'Scripting Egypt', 126.
116 Quoted in Thompson, *Wonderful things, Vol* 2, 59.
117 Deirdre Lawrence, 'Wilbour: One man's obsession with Egypt,' *Brooklyn Museum Blog*: www.brooklynmuseum.org/community/blogosphere/2010/03/22/wilbour-one-mans-obsession-with-egypt/ (accessed 7 June 2021).

118 'America's new Diplomatic Agent and Consul General', *The Egyptian Gazette* (2 September 1893): 2.
119 Allen, 'Tycoons on the Nile', 71.
120 See Sheppard, 'Trying desperately to make myself an Egyptologist'.
121 JHB to family, 1 November 1894, JHB Papers, Box 4. JHB Correspondence.
122 Janssen, *The first hundred years*, 12–13; Sheppard, 'Margaret Alice Murray and Archaeological Training', 113–28.
123 JHB to family, 1 November 1894, JHB Papers, Box 4. JHB Correspondence.
124 C. Breasted, *Pioneer to the past*, 64.
125 See conversion Table 0.1. This would be equivalent to around £11,600 or $20,495 today.
126 See conversion Table 0.1. This would be the equivalent of around £701 or $1,092 today.
127 C. Breasted, *Pioneer to the past*, 75.
128 *Ibid.*
129 Petrie had believed these remains were from what he called the New Race, but here I use the term pre-dynastic, meaning the period in ancient Egypt before the historically recognised dynasties (c. 3000 BCE). See Kathleen Sheppard, 'Flinders Petrie and Eugenics at UCL', *Bulletin of the History of Archaeology* 20:1 (2010): 16–29 and Debbie Challis, *The archaeology of race: The eugenic ideas of Francis Galton and Flinders Petrie* (London: Bloomsbury, 2013).
130 W. M. Flinders Petrie and J. E. Quibell, *Naqada and Ballas 1895* (London: Bernard Quaritch, 1896), ix.
131 This was the second season he was working with his now-famous Quftis. See Stephen Quirke, *Hidden hands: Egyptian workforces in Petrie excavation archives, 1880–1924* (London: Bloomsbury, 2010); Wendy Doyon, 'The history of archaeology through the eyes of Egyptians', in *Unmasking ideology in Imperial and colonial archaeology*, eds Bonnie Effros and Guolong Lai (Cotsen Institute of Archaeology Press, 2018). Allison Mickel also addresses issues of the erasure and silencing of locally hired labourers (*Why those who shovel are silent: A history of local archaeological knowledge and labor* [Louisville: University Press of Colorado, 2021]).
132 Petrie and Quibell, *Naqada and Ballas*, viii–ix.
133 See Alice Stevenson, '"To my wife, on whose toil most of my work has depended": women on excavation', in *Petrie Museum of Egyptian Archaeology: Characters and collections*, ed. Alice Stevenson (London: UCL Press, 2015), 102–5.
134 C. Breasted, *Pioneer to the past*, 76.
135 *Ibid.*
136 See Reid, *Contesting Antiquity*, 19–29; Allen, 'Tycoons on the Nile'; and Amara Thornton, '"... a certain faculty for extricating cash': Collective sponsorship in late 19th and early 20th century British archaeology', *Present Pasts* 5:1 (2013): 1–12.
137 C. Breasted, *Pioneer to the past*, 77.
138 *Ibid.*
139 Abt, *American Egyptologist*, 46–7; 281–301.

140 C. Breasted, *Pioneer to the past*, 146; Gaston Maspero to JHB, 16 October 1905, JHB Office Files, 1905, JHB Correspondence.
141 Abt, *American Egyptologist*, 126–8.
142 Andrews' Diary, 24 November 1905.
143 JHB to Hattie Breasted, 26 November 1905, James Henry Breasted to Hattie Breasted 1905, JHB Correspondence.
144 Abt, *American Egyptologist*, 130, calls the boat the *Mary Louise*, but in Breasted's letters to his mother, Breasted calls it the *Abu Simbel*. It was common to rename a rented boat.
145 JHB to Hattie Breasted, 30 December 1905, James Henry Breasted to Hattie Breasted 1905, JHB Correspondence.
146 *Ibid.*
147 See Abt, *American Egyptologist*, 131–2, and 'The Epigraphic Survey: The "Chicago House Method"'. For more analysis on photography in Egyptology, Riggs' *Photographing Tutankhamun* is indispensable. However, Riggs did not discuss Breasted's experience with photography.
148 Abt, *American Egyptologist,* 139; C. Breasted, *Pioneer to the past*, 168–9.
149 Abt, *American Egyptologist*, 156.
150 *Ibid.*, 156–7. See conversion Table 0.1. This is the equivalent of $394,500 today.
151 Abt, *American Egyptologist*, 211–13.
152 *Ibid.*
153 'Breasted sails with John D. Jr. on Egypt Tour', *Chicago Tribune* (3 January 1929).
154 Adams, *The millionaire and the mummies*.
155 Andrews' Diary, 12 December 1889.
156 Andrews' Diary, 13 December 1889.
157 The passenger list for the Sethi for a trip leaving 7 February 1890 had twenty-six guests, not including the medical adviser, manager, and crew (Cook archives, Black Box 2).
158 Andrews' Diary, 25 December 1889.
159 Andrews' Diary, 29 December 1889.
160 Andrews' Diary, 30 December 1889.
161 Andrews' Diary, 29 December 1889.
162 Andrews' Diary, 3 February 1890. Mohassib was so important to the discipline of Egyptology and collections around the world that Newberry memorialised him upon his death (Percy Newberry, 'Notes and News', *Journal of Egyptian Archaeology* 14:1/2 (May 1928): 184). Further, he has an entry in Bierbrier's *Who was who in Egyptology*, 376–7. Charles Wilbour wrote to his family a lot about Mohassib. Once, saying he 'looked over Mohammed Mohassib's *Klekshin* as he calls it, again … I am getting a lot of stuff' (Capart, *Travels in Egypt*, 292).
163 Adams, *The millionaire and the mummies*, 171.
164 *Ibid.*
165 Andrews' Diary, 6 March 1908.

166 Andrews' Diary, 27 December 1896.
167 *Ibid.*
168 Andrews' Diary, 9 January 1897.
169 Sarah Ketchley, 'Witnessing the 'Golden Age': The Diaries of Mrs. Emma B. Andrews', *KMT* (December 2020), 39.
170 Quoted in Adams, *The millionaire and the mummies*, 270.
171 *Ibid.*, 48–9.
172 *Ibid.*, 50.
173 Thompson, *Wonderful things, Vol 1.*, 150–1. He also notes that as late as 'the early twenty-first century, the Egyptologist Kent Weeks had his large dahabiyeh, the *Kingfisher*, home-ported at Luxor and occasionally sailed around Upper Egypt on it' (151).
174 See Carruthers, 'Credibility, civility, and the archaeological dig house'.
175 Albert M. Lythgoe, 'Note', in Andrews' Diary, n.p.

4

Luxor: archaeology with Thomas Cook

When approaching Luxor on the river, by steamer or dahabeah, guidebooks informed travellers that Karnak Temple would appear, slowly revealed by the trees: 'first the great obelisk, and the pylons ... half-concealed by palm-trees'.[1] Once the boats were even with Karnak, travellers would be able to see on the West Bank 'first the Colossi of Memnon and afterwards the Ramesseum'.[2] Many travellers were full of anticipation coming into the city. They had not seen a bigger town since Abydos or Girgeh, and they may have been awaiting mail since leaving Cairo more than a week earlier. Captain and crew would be singing and celebrating their arrival in Luxor, excited to see family or at least have a little break from the Westerners on their boat. But the travellers were looking for the ruins, for the town, for some sign of civilisation (Figure 4.1). Amelia Edwards recalled coming into Luxor in 1874:

> The river widens away before us; the flats are green on either side; the mountains are pierced with terraces of rock-cut tombs ... but [we see] nothing that looks like a Temple, nothing to indicate that we are already within recognizable distance of the grandest ruins in the world. Presently, however, as the boat goes on, a massive, windowless structure which looks (Heaven preserve us!) just like a brand-new fort or prison, towers up above the palm-groves to the left. This, we are told, is one of the propylons of Karnak; while a few white-washed huts and a crowd of masts now coming into sight a mile or so higher up, mark the position of Luxor ... we make our triumphant entry into Luxor. ... The assembled Dahabeeyahs dozing with folded sails, like sea-birds asleep, are roused to spasmodic activity. Flags are lowered; guns are fired; all Luxor is startled from its midday siesta.[3]

In early 1890, Emma Andrews, travelling with Theodore Davis and friends, described their arrival much the same way Edwards had:

> Flags were fluttering too from several dahabeahs, and a Cook's steamer. As we passed a dahabeah with the stars and stripes at the mast, M[ohammed].S[alah]. fired two salutes, which were answered by her, and a salute from the American

Figure 4.1 Vue general de Louqsor, dahabeah moored with temple ruins of ancient Thebes in background, Luxor, Egypt, c. 1867–99.

> Consulate which we returned, so that we made a proper sensation as we moved to our mooring ground, just beyond the Luxor Hotel.[4]

By 1892, there were telegraph posts and wires in Luxor, which stood in sharp contrast to the ancient remains. By this time, shooting guns as one arrived by dahabeah was a practice few adhered to, but raising one's country's flag, and seeing the same on others' boats, was always a welcome sight. Most boats would come all the way into town and dock at the quay, right next to the Luxor Hotel, like Andrews and Davis did the first time they arrived in January 1890. By 1907, the quay was between Luxor Hotel and the new Winter Palace Hotel, and according to the 1908 version of Baedeker's *Egypt*, the new Winter Palace Hotel was visible and 'conspicuous' upon arrival (Figure 4.2).[5] Travellers could not miss it.

By 1898, one could take the train all the way into Luxor if one wished. And a few years later, the railway ran all the way to Aswan. Boarding the train in Cairo, travellers could make the entire journey to Luxor in fourteen hours instead of fourteen days by dahabeah. Around 360 miles from Cairo, the views out the right side of the train included Dendera, just below Luxor,

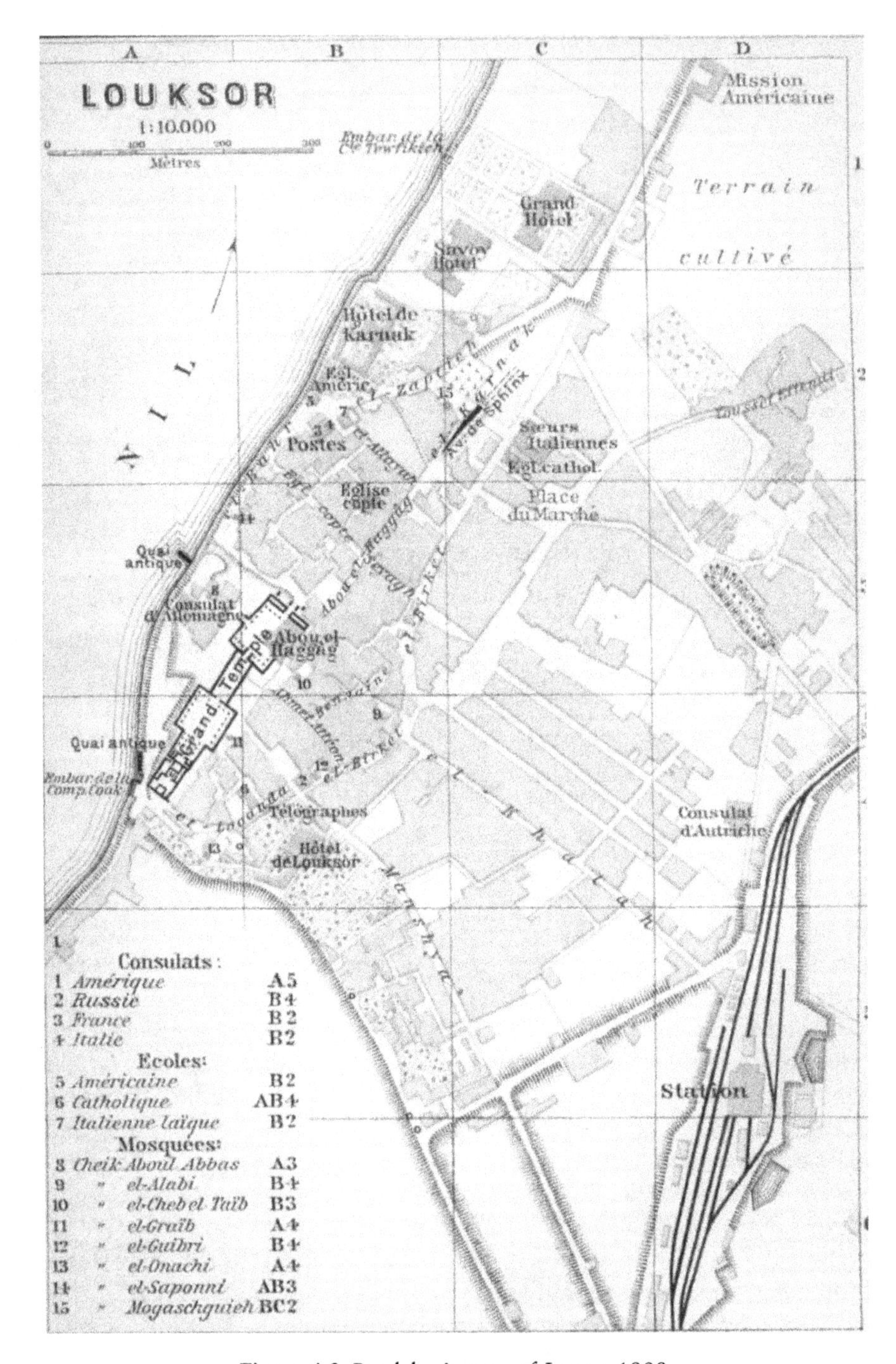

Figure 4.2 Baedeker's map of Luxor, 1908.

and of course the West Bank of the Nile in the distance and, closer, the Temple of Karnak appeared in the window; 418 miles from Cairo, the train pulled to a stop at the Luxor Station, just southeast of town. Arrival at this train station was not unlike arrival in Cairo from Alexandria – with donkeys, hotel porters, people selling antikas and other wares – just a little less crowded.[6]

Hotels in Luxor, like those in Cairo, were used as informal meeting spaces among friends and colleagues and for meeting crews as they came into the country. They were also used by new and aspiring Egyptologists who were looking for patrons or excavators. Hotels therefore played a central role in the creation of the intellectual landscape here, in that they created the roster of scholars, artists, and archaeologists who were allowed to participate in the discipline. Obviously, membership in this exclusive club was restricted to those who had enough time and money to get to Luxor to work in the first place. Moreover, the two main Western hotels in Luxor – the Luxor Hotel and the Winter Palace Hotel – were used not only to exclude Egyptians from participating in the construction of their own historical narrative, but also to keep archaeologists and their entourages disentangled from as many outside interactions as possible. Tourists, journalists, and potential crew members were usually milling about the gardens, ballrooms, and public lounges in the hotels, always willing to ask a question, get the next scoop about a new discovery, or find a job. But in hotels, archaeologists – needing peace and quiet to talk, think, or write – could escape the masses in the relative privacy of their rooms and suites. In these ways, hotels in Luxor were clearly, like those in Cairo, colonial spaces in which Western Egyptologists built and manipulated the cognitive landscape of the discipline by using the spaces as exclusionary truth spots and scientific institutions.

In contrast to the hotel spaces in Cairo, though, hotels in Luxor were well used by excavators during active seasons in the field, so they often operated as semi-public dig houses. Egyptologists continued to meet friends and colleagues here, but they would also perform much of the heavy intellectual work in the comforts of the public hotel lounge or their private room. It was in these places that they would discuss the day's labours, often store artefacts in their hotel rooms, and perform administrative tasks like writing letters, balancing budgets, and training crew members.[7] In these ways, hotels in Luxor were even more powerful than those in Cairo in terms of colonial knowledge building. Most Egyptians were not allowed into hotels unless they worked in them or held seats of social or political power, so outside of the excavation site where their local knowledge and expertise were invaluable to European and American Egyptologists and excavators, in hotels their voices could not be heard and their faces could not be seen. With few exceptions, they were denied access to the administrative and idea-making part of the excavations, yet they were depended upon for crucial labour within hotels just like they were depended upon for crucial labour at sites and on boats. As we will see, in Luxor especially, hotels were places of exploited labour and became tense places in which both sides fought for control of the colonial past, present, and future.[8] Again, Europeans and Americans were dependent on crucial Egyptian expertise while at the same time they erased Egyptians from the story of Egyptology and travel in Egypt.[9]

In this chapter, I continue to discuss dahabeahs as their own form of semi-private housing, not unlike a dig house or hotel in this particular town. These private boats, having made the long journey up the Nile to Luxor, continued to house travellers when they were moored. Sometimes these travellers would stay in a hotel for a few days or for the length of their stay, only to get back on board their dahabeah when it was time to leave. The people I discuss below worked on site, kept artefacts on their private boats, and often had their crew working and sometimes sleeping on board. While moored in Luxor and other places, as I have shown, these boats were public work place institutions, private domestic spaces, and everything in between.

Tourists came for the sights, sounds, and rest; Egyptologists were coming for much the same thing. Many came for work; they also used Luxor as a way-station, to have a drink, a tea, a meal, a meeting, then head back out to their dig site five to thirty or more miles away. Even if they were not staying long term in Luxor or had a dig house to use, such as the eventual Metropolitan House, Chicago House, and Carter's and Davis' dig houses, as an archaeologist they would need to stock up on supplies in town. However, they would frequently meet at a hotel if they were not staying in one. The story in this chapter is concerned with both excavation work on site, which has been thoroughly sited as a crucial space for archaeology and Egyptology, and hotels as central sites for this work. What scholars did within these hotels as guests, artists, archaeologists, and tourists meant that these hotels were centres of the creation of Egyptology at the turn of the century. This chapter brings together all of the elements of the spaces discussed in the chapters above.

Approaching Luxor as a tourist–archaeologist

In 1926, Thomas Cook & Son published a small pamphlet about the history of the company in Egypt. They opened with the bold claim that 'No country in the world has benefited more than Egypt from the enterprise of Messrs. Thos. Cook & Son'.[10] They were not far off in that claim. In almost sixty years, as this pamphlet outlined, Cook's had established electricity all over

Cairo, a regular mail service up and down the river, engineering works at the Bulaq, and built hundreds of boats for themselves and the Egyptian Government. Their most important achievement, they argued, was that they brought Western visitors to Egypt so that these industries would be necessary (instead of seen as luxuries for locals), and therefore the services could flourish. Cook himself (rather, the company) argued that because of Cook, 'came the building of luxurious hotels, the clearing of temples, and all that has turned Egypt into the most delightful of winter resorts, and the creation of a "tourist industry" which, after cotton, has become the dominating feature of Egyptian economic life'.[11] They said nothing about what this did to Egyptian lives, as a matter of course, but Cook & Son thought it was all for the better.[12]

In addition to building infrastructure, Luxor, some historians have argued, was a town that Cook built. While this chapter is not specifically about Cook, it is difficult to talk about travelling to and in Luxor without Thomas Cook and the company that was his legacy. To say that Cook built Luxor is only partly accurate; not only would that leave out the thriving community that had been there for millennia, but it would also leave out the fact that European Egyptologists were coming to Luxor before the mid-nineteenth century due to finds on the East and West banks of the Nile. The temples on the East Bank as well as the vast and rich mortuary remains on the West Bank – both of which are still producing material through excavation – brought far more tourists to town than the fresh, dry, desert air. Yet, as with Cairo, some people came for the fresh air, only to stay for the Egyptology.

Baedeker did not publish many editions of the Upper Egypt volume of their Egyptian guidebook. Published separately as the second volume of a whole guidebook, the standalone book contained, as its subtitle described in detail: 'Upper Egypt, with Nubia as far as the Second Cataract and the Western Oasis'.[13] After some issues with author contracts and costs, it was published on its own in 1892, after the 1885 edition of Lower Egypt (and the Sinai Peninsula) had gone into its second edition.[14] An edition for Upper Egypt came to be necessary as more and more people were travelling to the First Cataract and beyond; 1898 saw the first combined edition of Baedeker's *Egypt*: it contained the entire journey travellers might make in Egypt, both Upper and Lower parts, as well as the Sinai Peninsula and into Nubia (Sudan).[15]

Luxor was, and remains, a smaller municipality than Cairo and not everyone who went to Cairo for the winter months made it to Luxor. Because relatively few people travelled to Egypt, and even fewer made it up the river, Luxor was a privileged place from the start. Because of this, everyone knew who was coming to Luxor, on which vessel, when, and most importantly, why they were coming.[16] Tourists were coming for clean air, to spend relaxing days on the river, visiting with friends and of course touring the temples, tombs, and bazaars. Even with relatively few visitors compared to those who travelled to Cairo, the village of Luxor became a boom town during the travel season of October to April. During these cooler months, it became overcrowded with independent tourists and those who booked guided tours through Cook, Gaze, or other smaller companies. Pierre Loti, visiting in 1910, described a Cook's tourist steamer in Luxor as: 'an enormous black pontoon, which spoils the whole scene by its presence and its great advertising inscription: "Thomas Cook & Son (Egypt Ltd.)"'.[17] He wondered aloud to his reader: 'How shall I find a quiet place for my dahabiya, where the functionaries of Messers Cook will not come to disturb me?'[18] Loti and others would be hard-pressed to find such a place.

There were only a few hotels in Luxor until the 1920s or so, probably adding to the overcrowded feeling of the town, with too few vacant rooms. In fact, until 1877, when the Luxor Hotel was built, those who did not have a dahabeah or steamer berth would have to sleep in a chamber at the Karnak Temple, or in the home of a Luxor resident.[19] The Luxor and Karnak hotels, originally built and owned by Thomas Cook, were the best and most popular European hotels in town until 1907. January of that year marked the opening of the Winter Palace Hotel. This is the only one of these three early Western hotels still functioning as a hotel. Each of these hotels played a role, some more significant than others, in the history and practice of knowledge creation in Egyptology.

From 1877 to 1907, there were a few smaller hotels such as the Savoy and the Grand Hotel Tewfikieh (later just the Grand Hotel). The Grand Hotel was owned by Henry Gaze, Cook's biggest Nile travel competitor, but since Gaze was out of business by 1903, the hotel did not last long. According to the 1902 Baedeker's *Egypt* map of Luxor, the Grand and the Savoy, two of the first hotels one encountered coming upriver into Luxor, had massive gardens that swept from their respective fronts all the way down to the Nile's East Bank. The next hotel visitors would see as they sailed into town was the Karnak Hotel. Its façade was right on the river, and its gardens were behind the hotel, away from the river. Built by Thomas Cook a few years after the Luxor as a sort of Luxor Hotel annex for Cook's tourists and Egyptologists, the Karnak was right next to the Savoy. It was a bit cheaper than the Luxor Hotel, as it was not as nice and only accommodated about sixty people as overflow from the Luxor. The Karnak closed for good after the First World War, as a number of hotels in Egypt did, due to lack of business combined with the British military having commandeered them. If archaeologists or Egyptologists were going to meet anywhere between 1877 and 1907, Luxor Hotel was the place.

Luxor Hotel

Built adjacent to the Luxor Temple, the Luxor Hotel opened in 1877. It was the first Thomas Cook & Son-built and owned hotel anywhere in the world.[20] Inside, visitors would find Egyptian-style decorations in the music, smoking, billiard, and drawing rooms. The gardens were large and luxurious, with palm trees, aromatic flowers, and collected antiquities displayed around the grounds.[21] Despite the opulent setting, it took a while for the hotel to be brought up to the standards that tourists expected of Cook. Dozens of customer complaints about bad food, high prices, dingy rooms, and more, pushed Cook to make improvements to the hotel. They had done so many upgrades in such a short time that by 1880, Murray's *Handbook* described the hotel as excellent. This rating, and Cook's steamers, brought even more tourists to the town. Having a Western-run hotel in Luxor, finally, meant that tourists could spend more leisure time seeing the sights. They could travel up the Nile on one steamer, stay in Luxor for an extended period, and return to Cairo on another steamer, boat, or train.[22] This was a brilliant way for Cook not only to make money, but also to increase tourism in and around the area of Luxor.

By early 1889, however, residents in the Luxor area complained to Cook's offices that 'travellers still amuse themselves by trespassing on growing crops & generally taking with them a swarm of Arab boys who do considerable damage to the crops'.[23] John Cook wrote to his son, Bert, that 'I think we are bound to take steps to try to prevent such being done if possible'.[24] John himself made a sign to be posted on each steamer, in each hotel, and in all Cook's offices exhorting people to stay away from farm land and be respectful of the local population. He had the sign translated into French and German, as well as English.[25] This was a wise move on John's part as just a few years later, in 1891, Flinders Petrie discovered a beautifully preserved painted floor at Amarna. Although Petrie built a structure over the floor to protect it on site, by 1911, frustrated and angry farmers destroyed the entire floor to keep people away, because tourists were flocking to the site through their cultivation, destroying crops in the process.[26]

There were also 'very loud & strong complaints against the Hotel' by paying customers in the 1888–89 season.[27] John recounted these in a letter to Bert. Customers thought the food was monotonous and bad, and, despite its terrible quality, there was never enough of it. The lounge chairs were uncomfortable 'straight-backed chairs' with

> scarcely any easy chairs except the [steamer] deck chairs taken by the passgrs [sic]. They also complained that there is not the slightest attempt at any amusement … that there is nothing whatever about the Hotel to interest or entertain them & I find that quite a number are going to shorten their stay for these reasons.[28]

Shortening stays would mean loss of income for the company that season, and possibly the loss of future customers. Uncomfortable in the hotel, some people left Luxor all together, moving on to Aswan or simply heading back to Cairo, then home. John wondered if others went to the Karnak Hotel, as a 'plea of cheapness – wanting lower rates but they themselves say it is because they did not get proper attention & comfort'.[29] John wrote to Bert that the complaints were 'annoying to me in more senses than one', mainly because 'it is understood by everybody that the [Luxor] Hotel is as much my property as the steamers & therefore we get the blame for this dissatisfaction & bad management'.[30] The company owned the hotel but did not run it (and by the end of 1889, Cook did not own it, either). Cook's goal was customer satisfaction, and he was not, by and large, happy with the results at Luxor.

By the following tourist season in 1890, it seems many of the hotel's problems were fixed. In her diary from that year, Emma Andrews wrote about her group's quick visit to the Luxor Hotel, mainly 'to see how far it bore out the praises lavished upon it by Cook's pamphlets'.[31] She must have thought it did so satisfactorily, because she continued, describing the hotel and her impressions:

> The entrance gate is near the landing and we walked through a pleasant avenue of Scent trees, their pretty little tufted yellow flowers making the air fragrant with their delicate perfume. The house looked very attractive and especially the newer portion now occupied by the Grand Duke Peter of Russia and his suite – and he sat reading on one of the low balconies. The house is built of Nile bricks of mud, with enormously thick walls – and looked comfortable enough, though the rather English style of everything contrasted strangely with the crude style of building. We sat down in the shade of the verandah to read our letters. I saw several sad face [sic] invalids, walking slowly about the grounds – and a hideous stout old hag of an Englishwoman, enveloped in furs, with gold chains and beads and bracelets and rings, sitting in a wheel chair. She was such a *type*, that I wanted to sit down in front of her and think of her, and talk to her. On one of the low balustrades, was a huge pelican – a magnificent creature, quite tame, and most affectionately devoted to a young pointer dog, who returned the attentions most demonstrably, and it was very comical.[32]

The hotel must have been upgraded even further by then. Described by Baedeker's *Egypt* two years after Andrews and Davis had made their first stop, the hotel had 'a fine large garden in which several interesting stones are placed'.[33] There were, in fact, several statues and artefacts throughout the gardens of the hotel, and at one point the hotel had reused door jambs from a near-by tomb.[34] Travellers also learned the hotel had discounts for 'Egyptologists and those making a stay of some time. Pension includes morning

Figure 4.3 Luxor Hotel, front lawn and façade, c. 1880–89. The sign for Cook's Office is visible.

coffee, lunch about noon, supplied also to those making excursions, and a substantial dinner about 6 p.m. The rooms are clean, but not luxurious.'[35] Being a long-term resort hotel, the Luxor had a resident English doctor, Dr Longmore, who also worked at the local hospital.[36] There was also a Cook's office attached to the hotel (Figure 4.3).

The façade of this hotel still stands in Luxor, just behind Luxor Temple and McDonald's restaurant, awaiting rebuilding. Deserted now, it used to be a very busy place – not only for tourists but, as Baedeker's revealed, for Egyptologists to stay close to sites on the East Bank. Or for them to have a place to stay when they were coming into Luxor from sites further afield for a few days. There were plenty of important Egyptologists who stayed at the Luxor, creating there a comfortable space to withdraw from the heat and the site, to work in privacy and peace, and to connect with their fellow travellers. Maggie Benson, her brother Fred, and her partner Janet Gourlay all stayed there as they excavated the Temple of Mut from 1895 to 1897. Harold Jones, the lung-tripper artist who first established himself in Cairo, also stayed at the Luxor Hotel when he was working with Andrews and Davis. What is important is not only what these travellers said about the hotel, but also the fact that they stayed there at all and what their activity in the space means for the network of people present in Luxor at this time. The work they did, the relationships they made, and the way in which this hotel supported field Egyptology and Egyptologists made it a centre of scientific work and networking for these archaeologists. The Luxor Hotel was a hub of a wide network that changed each year depending on who arrived in Luxor, but Egyptians were never included.

The Temple of Mut at the Luxor Hotel

Margaret (Maggie) Benson arrived in Alexandria, Egypt with her brother Fred in January of 1894, searching, like so many others, for a treatment or cure in the warm, dry air of southern Egypt. They were headed up the Nile on a Cook's steamer, and Benson was trying to relieve symptoms of her physical illness and mental and emotional strain.[37] She suffered from rheumatism and arthritis as well as pulmonary ailments associated with these auto-immune diseases for most of her life. The first time she and Fred were in Egypt, they stayed in Luxor for almost two months, and made their home in the Luxor Hotel. They would continue to do so as they returned for excavations. Benson enjoyed the hotel and told her mother how distractingly beautiful the surroundings were:

> Such a garden to write in, that it is almost impossible to write.–One thing that keeps attracting one's eyes is a red bougainvillia [sic] with sun on it against a background of pale blue sky and greyish palm-trees, it is wonderful – and the birds are distracting.[38]

It was this garden, and the sense it gave her that she could 'think about nothing except what one wants to do', that brought her back the next year to excavate.[39]

Benson herself is an enigmatic character in the history of Egyptology. Like others, mostly men, who started their careers in Egyptology upon arriving in Egypt, Benson had no training in excavation or Egyptology. She had no proficiency in Arabic or hieroglyphs. She did, however, have excellent educational training. She had attended Lady Margaret Hall, Oxford in 1883 and excelled there. In 1886 she earned First Class honours (but no degree) in the Women's Honour School of Philosophy. As with many women working in the field in the late nineteenth century, if Benson had been a man, it is possible her life, and her impact on Egyptology, would have been seen differently. Her brother Arthur wrote of her:

> My sister had, I do not doubt, a great intellect, both clear and profound ... But as things were, these pursuits were to her private interests, which she followed for her own delight, when leisure permitted. She had a great faculty of working on quietly and alone, and was not in the least dependent on intellectual sympathy and companionship ... She seldom talked about her work, never appeared burdened by it or absorbed in it, never brandished it in the face of an intruder.[40]

Fred was in and out of Luxor, helping David Hogarth and travelling on his own, but he supported his sister's excavation efforts. But Maggie, feeling the benefits of the hotel garden and the climate of Southern Egypt, worked out a plan to return in 1895 to excavate in Karnak Temple, in an area known as the Temple of Mut. Mut was the consort of Amun, whose main temple was Karnak, and together with Amun and their son Khonsu, Mut was one of the Theban triad. Auguste Mariette, with Karl Lepsius, had cleared part of Mut's Temple area in the 1840s, and they had also created a map of the temple, which appeared in Mariette's 1875 book about Karnak.[41] Charles Wilbour had also improved on Mariette's map in the 1880s, when he was travelling with Maspero. All of this seemingly complete work is why, in their first season, which ran five weeks from 1 January 1895, the group had not expected to find anything. At the same time, the lack of expectation is likely why they were given permission to excavate there, and they were told as much before they began.[42]

Their first application for permission had been refused, but the Swiss Egyptologist Edouard Naville promptly wrote to Jacques de Morgan, then-head of the Department of Antiquities, in support of Benson's application; de Morgan granted the permission immediately. Despite the barriers she faced, she wrote to her mother as she made preparations for the first season that 'I find that I am beginning to be considered in the light of an Egyptologist'.[43] They started their work with 'four men and sixteen boys, an overseer, a night guardian, and a water-carrier'.[44] She also had the help of Egyptologist Percy Newberry – who was working on a survey of the Theban necropolis across the river – Naville, Flinders Petrie, and David Hogarth.

As they got to work, Benson did not think they would find much outside of 'little things, only walls, bases of pillars, and possibly Cat-Statues'.[45]

Surprisingly, in that short season they found dozens of statues, pieces of statues, coins, and other small artefacts. The best artefact from the year was a black granite statue that brought attention from Georges Daressy, then the Inspector of Upper Egypt.[46] The statue was of a young scribe, Amenemhat, from the reign of Amenhotep II, 'and was almost perfect'.[47] After some discussion with the Egyptian sub-Inspector, Daressy told Benson and her brother that they could remove the statue to their rooms at the Luxor Hotel for safe keeping. First, it sat outside the hotel to be photographed, then, Benson wrote to her father, 'I am going to have it in my room'.[48] The statue eventually went to the Museum at Giza, in the museum boat, and Benson got a cast of the statue to keep.[49]

Their second season started well in January 1896, with more staff than the previous year so they could cover more ground and even work simultaneously at two separate pits. This was the season that Lady Jane Lindsay introduced Benson to Janet Gourlay. Lindsay, a friend of the Benson family, was visiting Luxor that year and also helping with some excavation work. She wrote to Benson's mother that she found Maggie doing much better:

> How I have wished you could have stepped into my shoes when they walked about Luxor Hotel Garden in search of Maggie last Tuesday. She *is* looking so well – and is in highest spirits about her own health – I wish you could just see her, so full of vigour and quick movements.[50]

The Egyptian air and the distractingly beautiful gardens helped her to feel better – both physically and mentally. Once they met, Benson and Gourlay became fast friends, then lovers and partners for the rest of their lives.[51] By May of 1896, Benson wrote to her mother about Gourlay as she had never written about anyone else:

> how can I keep you up in this, for it changes to every day – oh, I *hope* you'll like her – you can't help it if you know her, but she is so horribly shy. She is only 33, but she makes me feel like a little girl sometimes – and you know I don't do that particularly easily ... She told me she hadn't ever talked so much to any one before. Oh, Mother, it's so odd to me to make a friendship like this – generally there has been something in the way – mostly I've not been sure of the other person, and generally I've had a radical element of distrust. But here one can't help trusting her *absolutely*, and it's only myself I distrust. She is so much bigger, and so much finer and more delicate in mind than most women.[52]

During the highly productive 1896 season, the team were able to update Mariette's map of the Mut precinct of the temple, and they found statuary quickly upon beginning their work on 30 January 1896. One of the pieces they found was the upper part of a statue of what turned out to be Tutankhamun, but it had been buried face down in the temple fill. Benson wrote of the careful work they did on this statue that, '[i]t tries the patience of an excavator to work slowly at a statue which is lying flat on its face, so that the most important point cannot be determined until the whole thing is free'.[53] But free it they did, and they were able to set the statue back upright in the temple.

Toward the end of the excavation that year, and a few days after finding almost a dozen statues and pieces of statues in granite and alabaster, there was an incident involving criticism of her methods that shook Benson. She wrote to her mother that one morning Georges LeGrain, the Inspector at Luxor at the time (and until 1917),

> descended on us ... and upbraided me with having taken all the things to the hotel, saying he should take all the little statues to his store-house. I very nearly wept, and called Fred, who was slightly rude. M. Legrain became much more polite and finally said if we chose to take the whole responsibility of their safety, we could take them back [to the hotel], and he would write to DeMorgan. So we did.[54]

What Benson, Gourlay, and their crew were doing, however, was not so different from what other excavators had done. Petrie was known to put finds in boxes and then lay his mattress on top of those boxes to form a bedframe on which he would sleep. LeGrain's attack on Benson and Gourlay can be seen as an attack by chauvinistic officialdom: LeGrain thought of Benson as a woman who did not know what she was doing. Once her brother arrived, according to Benson, LeGrain's demeanour changed, the presence of a man (and one with archaeological experience) reassured LeGrain that the work was being properly undertaken.

In the end, after only about five weeks of digging in 1896, they had found dozens more pieces of statues and two nearly complete statues – one of Senenmut and one of Bak-en-Khonsu.[55] They were allowed to keep a statue of Ramesses II and heads of Ramesses III and an unnamed god. Regardless of the fact that Benson had no training outside of her brother's early guidance, she was a successful excavator. Furthermore, Gourlay had been a student of Petrie's at UCL in 1893, so we can reasonably assume that she was trained in his excavation methods: quick and thorough – and cheap.

Her final season began even more auspiciously than the previous two. Thanks to some extra money from an anonymous donor and from Gourlay's father, Benson and Gourlay could afford to hire a larger group of workmen and boys. Much of Benson's family joined the excavations that year, looking for some respite following the death of her father in October 1896.[56] They began work on 10 January and immediately found more statuary in the same place their 1896 dig season ended.[57] They found in the morning of that first day two large statues of black granite. One piece belonged to a statue fragment from the 1896 season, and another was the head and shoulders of a priest from Thebes.[58] Also on that first day they found fourteen pieces of statues and some of them with heads and faces intact. They found dolls, pieces of roofing and flooring stones, and most importantly they uncovered a number of column bases with which they traced a more accurate map than Mariette had been able to produce.[59]

Only a few miles away from where the Temple of Mut was located at Karnak, the Luxor Hotel itself was the base of the excavation crew for those three seasons. As such, the hotel was the space in which Benson would meet with her brother, Gourlay, and other archaeologists working on the excavation to discuss the work, store and catalogue the artefacts, and do all the excavation administration. She would write letters, make notes, figure budgets, and ready herself for the choosing of workmen upon arriving at the site. It was here that she also socialised with tourists, played games with her family and friends, and welcomed her mother, her mother's partner Lucy, and her other brother, Hugh, when they joined Fred, Janet, and Margaret in the third season in 1897. It is possible that during this time Andrews and Davis may have run into the Benson crew at the hotel, but we see no evidence of it in the letters from Benson to her family or in the Andrews diaries.

During this time, Benson also battled depression, so she often took to convalescing in her bed or the gardens at the Luxor Hotel. Even though her mental health was getting better thanks to the weather and her relationship with Gourlay, her physical health got worse. She was treated for a near fatal case of pleurisy toward the end of her last excavation season in 1897, followed by a heart attack. For those who develop pleurisy, the lining on the outside of the lungs can become so inflamed that it is almost impossible to breathe due to pain. Fluid can build up between the lungs and the lining of the lungs causing shortness of breath, and sometimes partial or complete lung collapse. Benson's condition was so serious that it was necessary to insert a needle and remove the fluid around her lungs so she could breathe. This operation was performed by an 'able physician', probably Dr Longmore, in her hotel room.[60] The procedure today is invasive and is usually done in a sterile operating room with anaesthetic and an ultrasound for guidance. Longmore would have had the needle, but no ultrasound to guide him, and there was no real local anaesthetic option to speak of. It would have been difficult, to say the least, but many medical procedures were done at home in the nineteenth century. After the procedure, and possibly related to it, she suffered a heart attack but recovered in about a week and the group made their way home.

When the two women were apart at home in England, they wrote passionate letters back and forth to each other. Her fervent love for Gourlay came through in these letters. Over the summer of 1896 Benson missed Gourlay and told her so, writing 'I had been dreaming of you – as usual unsatisfactorily – namely that we had gone to Egypt for a week only, and nonsense of that kind.'[61] Later, still lonely and wishing for Gourlay's company, Benson closed another letter: 'Oh dearest, I wish I knew the Gaelic language, for I believe you are able to say all sorts of affectionate things in it which English can't express. I do want you in bodily presence very badly, my dearest, Yours, M. B.'[62]

Benson's relationship with Gourlay was passionate and loving and should be seen as a scientific partnership that benefitted both women (Figure 4.4). Much like we have seen in heterosexual relationships, and will continue to see with Emma Andrews and Theodore Davis, in order to work in more isolated places like Egypt, women needed assistants, or to be assistants, to help them in a number of ways. These partnerships were expected of educated

Figure 4.4 Maggie and Nettie, c. 1906.

English women who travelled abroad, and for archaeologists, these relationships often operated very much like heterosexual marriages by giving women the freedom to travel where they needed to go without judgement or hindrance, taking along the support they needed in remote locations, and, importantly, the power to do their work. In Benson's and Gourlay's case, two educated women could work without needing a man, in a period when it was almost impossible for a woman to do so on her own.[63]

In the usually masculine and muscular history of Egyptology, there is a distinct erasure of queer partnerships in favour of discussing the importance of straight marriages (and even affairs) as factors in professional success. However, especially for women in field sciences like Egyptology, being in a same-sex scientific and domestic partnership clearly made them more productive professionally and more secure socially than they would have been as single women. Because of Benson, Gourlay had the funds and permissions to excavate in Luxor. By working with Benson, she had the opportunity to put her training to work and publish important scholarship. Because of Gourlay, Benson had a trained Egyptologist working on site with her who would treat her as an equal, and not as an assistant or subordinate. They were a true scientific couple in the field, and their productivity together demonstrates that point.

Through her personal and professional relationships, Benson was able to accomplish a lot of work in just a short time. Most of this work, including the continuation of analysis and organisation of the finds, was done within the confines of the Luxor Hotel, making the hotel not only an important space for creating knowledge about the Temple of Mut, but also a truth spot which gave credibility to the work the crew did. They used the Luxor Hotel as both a field site and as a dig house from which to centre their legitimate and important work.

Over the course of three seasons, Benson, Gourlay, Fred Benson, and most of the Benson family, visited Luxor to help excavate, or at least to supervise. The Egyptian excavation crew cleared much of the temple itself of landfill, exposing foundation deposits, and finding remains and pieces of over 100 statues and countless other pieces of sculpture. They also reorganised the interior, attempting to restore items such as columns and some statuary to their original locations. Some historians write about Benson herself as an amateur who came to Egypt on a whim and decided to dig for fun, which is not untrue, but this puts her in the same category of plenty of others in this same period – she just happened to be a woman. Others are dismissive of her work as unprofessional and, while important, not *as* important as Mariette or LeGrain's work on the same areas. Fred had excavated with the British School at Athens from 1891 to 1895, and with David Hogarth in Alexandria in early 1895, thus lending some authority to their work. But the organisation and plan of the work, and the report that came from it, were efforts led by Maggie Benson, in conjunction with Janet Gourlay. In reality, Benson's work on the site was and continues to be of use to Egyptologists working in Karnak.[64] Based at the Luxor Hotel, their work was also a marked shift in the normal gender roles in Egyptological excavations, especially in this period.

Patronage at the Luxor Hotel

When Harold Jones arrived in Luxor for the first time, he also stayed at the Luxor Hotel. As we saw in the second chapter, Jones had arrived in Cairo in 1904 looking for relief in the dry air of Egypt from his tuberculosis symptoms, much as Benson and others had done and would continue do. He had initially travelled to Cairo to work with the John Garstang crew from Liverpool. Garstang had been a student of Flinders Petrie's in 1899 at Abydos and, by the time Jones began to work with him, Garstang was an honorary reader in Egyptian Archaeology at the University of Liverpool, the university with which his name would remain attached for the next century and more.[65] Garstang had a lot of active projects on different sites in Egypt when Jones joined him, digging at Beni Hasan, Esna, Naqada, Hierakonpolis, Edfu, and more, so he put Jones charge of supervising and recording the excavations at Esna and Naqada. Jones was an artist by talent and training, and he was not particularly prepared for the isolation he would experience on the sites Garstang assigned to him, so almost from the start he started looking for another position that would bring him closer to a town and surrounded by more people.[66]

While looking for a new job, Jones would go to Luxor for meetings with friends, acquaintances, and potential future patrons as often as he could. He understood the power of the networks within these spaces. Jones wrote home in January of 1905: 'arrived at Luxor about 12.30 & met Garstang at the Luxor Hotel'.[67] From there, they went to meet with Archibald Sayce, Garstang's friend and mentor who also travelled with Charles Wilbour on his dahabeah, and then they went to Thebes to meet James and Annie Quibell.[68] The following week, Jones wrote again saying that he would stay at the Luxor Hotel for the night, while gathering supplies for Garstang's excavation. He said that he had tea with 'two ladies whom I know here' then went to Karnak Temple to see Georges LeGrain (the same LeGrain from Benson's excavations). Neither man knew the other's language, so they spoke to each other in Arabic.[69]

A month later, Jones was writing home from Luxor about James Quibell's uncovering of a new tomb, under the sponsorship of Theodore Davis and

Emma Andrews. We know of the tomb as KV 46, the tomb of Yuya and Tjuyu, the parents of Queen Tiye who was the wife of Amenhotep III. Being near the Valley of the Kings, Valley of the Queens, and the Tombs of the Nobles, Luxor and its hotels were a perfect spot to hear about, and go and see, a great many sites during this period. However, the artist Jones went a step further and hoped for a job closer to Luxor, with people who might value his contributions, which he felt Garstang did not. He wrote home:

> There's been great doings here a Mr Davies [sic] has discovered a wonderful lot of things – a practically undisturbed tomb – you will see all about it in "Times" the day after you get this as Professor Sayce is writing the full account ... I've seen everything & was one of the favored few to go down the tomb & see everything in position. The Duke of Connaught & of Devonshire & Empress Eugenie were there too ... I am going to try to get the job of drawing the whole lot for Mr Davis account but they have an American artist – a Mr Lindon-Smith so possibly it will not come to much.[70]

Andrews, Davis, and their group were staying on their dahabeah at this time, and, as the previous chapter explained, it was a central, privileged place for Egyptological activity. Jones soon became a regular visitor to their boat, but at this point in time, he was staying and meeting people more at the Luxor Hotel than anywhere else. By 1906, Jones was complaining to his parents that Garstang was away too much and that he would rather work for Davis and Andrews.[71]

By early March 1906, after a few meetings both on the *Bedauin* and at the Luxor Hotel, Davis had finally made him an offer: 'Mr. Theodore Davis – the "American Millionaire" who found that wonderful tomb at Thebes last year has asked me if I'd call to take on his plates &c for his publications – as I would get about £15 a plate I believe – I am thinking of accepting – if Garstang will let me have a couple of weeks off to start in.'[72] He was optimistic about the prospect of working for Andrews and Davis because he could make £200 per season if he could complete the work.[73] Garstang must have given him the time, because just over a week later, Jones was installed at the Luxor Hotel, working on the artwork for Davis' and Andrews' excavation report. As soon as tourists left for the summer, though, Jones hoped 'the tombs will be cool in the hot weather. I expect I shall live in camp at Thebes with [Edward] Ayrton who is excavating for Davis – but if the weather is trying I shall stay at this hotel so to have decent meals & go off at sunrise with my lunch returning at sunset for dinner.'[74] He loved being in Luxor, staying at the Luxor Hotel, and the social and professional attention the situation afforded him. Like Benson and Gourlay, Jones was able to use the Luxor Hotel to prepare his professional activities as well as support his personal life.

Jones continued to work a little for Garstang in 1906, but by November of that year he had fully joined Davis, Andrews, and their crew. He had met with the Americans in Cairo before he went to Luxor in December of 1906.[75] By the time he got there to begin work, he found Ayrton, American George Reisner, and other Egyptologists already there. He wrote home about the work and his colleagues: 'we are a goodly quartet & time passes merrily & quickly & the paintings are getting on apace though they will take another ten days or so before I can get them all finished'.[76] Even though they were staying on the *Bedauin* for most of the season once Davis and Andrews arrived, the town of Luxor was eight miles from the Valley and it cost time and money to get there by donkey and then boat. Jones enjoyed the visits from tourists on site, and wrote to his parents that with all of their work in Ramesses VI's tomb (KV 9):

> we are persons of no little interest!! ... Davis will be here in a weeks' [sic] time I expect – so that too will enliven things somewhat as he dines up here with his sister (Mrs Andrews) & niece twice a week generally. In fact I would rather live here than anywhere else in Egypt almost ... [it is] pleasant & amongst civilized people whereas all our other camps have been far from hotel life & large tours.[77]

Jones very much enjoyed being around others – tourists, the others on the excavation crew, the other artists, and Andrews and Davis. Despite the fact that Andrews was, decidedly, not Davis' sister, it is interesting that Jones believed she was. It was either that, or he wrote home to his family that Andrews was Davis' sister so as not to scandalise his parents. Undoubtedly for Jones, dahabeah and hotel life meant social interaction, new friends, discussion, and a new and exciting career.

It is clear that once Jones began working permanently for Davis and Andrews, his life in Luxor got a lot more exciting. A typical day for him, as he described in a letter to his parents in February of 1907, looked a little like this:

> You can hardly imagine all this never having been here – to begin with the hotel Luxor is on the East bank of the Nile & Davis' dahabeah is the other side of the river the tombs of the kings being still further in the desert some eight miles on the other side too. When I go to Davis' to dine &c I am met by a boat with six [A]rab oarsmen & a cox & am rowed across the river being nearly half a mile across so it's no small matter [getting] about here, then one is so well known to all old comers that introduction to everybody are as common as can be as we are all more or less interesting folk! So we are always told.[78]

His job on that particular February day was to draw a shrine 'which was found covered with gold stucco'.[79] Then Davis and he went to the '[Ernest]

Gardner tent to tea to find a larger party expected'.[80] This party included, among others, Maspero and Naville. Walking back to the site with Davis after tea, the two met Inspector Arthur Weigall and Lindon Smith. The men talked shop about KV 55, the tomb they were working on at the time, and, he continued:

> still further gossip when 7 o'clock some came then I had to hurry off to dress to go to Davis' Dahabieh to dinner at 7:30 & there I stayed till 10 o'clock having a fine time looking over the things he has found – gold diadems, canopic jars with portraits of Queen Thiuy [sic] &c &c at 10 o'clock I had to call for Miss Hardy (Davis' niece) & take her to the boat & then I went to the Hotel to find a quiet corner & was at once joined by a Mr. Hart, his wife & nephew whom I had met at the tea fete, they engaged me till after 12 o'clock when I was glad to go to bed – there's one day & a typical one.[81]

His down time at the hotel was usually busy because people knew to find him there. The day he was writing this letter, in the dining rooms of the Luxor Hotel, he told his family that he had plans to go to Davis' boat for lunch, then back to the tombs, but his writing was interrupted twice. Once by Baron Rothschild, a British banker and politician likely in Egypt on holiday, and then 'by Howard Carter coming up & wanting me to go to his room in the Karnac [sic] Hotel which he uses as a studio'.[82] It is not clear if he went with Carter at this point, but we do know that the two men were friends and enjoyed sharing their art with each other. It is clear that, later that day, he was back at the tombs where he kept getting interrupted by tourists. Weigall, Lindon Smith, and their wives were all there, also. Jones' fondness for the Davis–Andrews group is clear, as he closes this particular letter, reassuring his family:

> Davis is a Welshman by origin & is *very* kind to me – he wants to find me a wealthy American heiress!! & advises me to hyphen my name so as to sound more important than plain Jones – his butlers name is Jones so they call me Titian as a distinction! Mrs Andrews – a relation of Mr Davis is very concerned because I sometimes cough. She's been dosing me with Scotts Emulsion! Till April I am keeping very fit & well – hard work seems to agree with me.[83]

Later that same month, much to his surprise, he was painting all day on the Davis dahabeah and dining with wealthy American banker and patron of the MMA in New York City, J. Pierpont Morgan in the evening.

Although Jones did not appear in Andrews' diaries until January of 1907, this is not a significant omission, as by this time the pair had been going to Egypt for over a decade and she did not record as much of their daily activities as she had done early on. As usual, her detailed record tells us a bit about Jones' work for their party. She described the reproduction in painting of artefacts coming out of what would become KV 55 and how he dined with them regularly on their boat.[84] When he was painting all day in February of 1907, the Andrews–Davis party were out on an excursion, but Maspero had come with a few others to see the KV 55 treasures, and she recorded that, 'Harold Jones showed the treasures'.[85] Her recollections reflected what Jones had said of the group – that they enjoyed having him as part of their crew. They also trusted him to be on their boat alone, with priceless artefacts to work with.

Both the boat and hotel were sites of connection, mentorship, and patronage which propelled Jones into the view of these powerful and wealthy patrons, but his talent got him the work. Having this patronage network willing to pay a substantial wage for good work was extremely important not just to Jones but also to other aspiring Egyptologists like him. Where Benson could support herself because of her parents' wealth and status, dozens of others could not. Jones and Carter needed public spaces in which to meet people like Davis and others so they could get a job. The Luxor Hotel did not just act as dig house, but was also a central network node in early twentieth-century Egyptology. The Luxor Hotel was the centre of much of the work the group did and the administrative meetings they held. In addition to meeting Jones there, Andrews met with Howard Carter in the Luxor Hotel in early February of 1907, with Davis, and their family, Nettie and Jean Hardy.[86] Not much is known about what they said, but since Carter was working with them at the time as an artist and draftsman, we can imagine it was plans for work, as well as some personal conversation.[87] To be clear, Andrews was a formidable archaeologist in her own right, just as much as Davis was. She paid for at least half of a number of the excavations, did all of the recording on site, and kept detailed lists of visitors to the site and more. She preferred to do a lot of her business at the Luxor, as well. With the examples of the Benson crew and Jones and Carter, we have seen that the Luxor Hotel was a crucial site for how Egyptology was done in Egypt.

While the Luxor Hotel was accepting guests until the 1980s, the opening of the Winter Palace Hotel caused a shift in the centre of work in Luxor in early 1907.

The Winter Palace Hotel

On 19 January 1907 the *Egyptian Gazette* was raving about the opening of the Winter Palace Hotel in Luxor. The new hotel opened to great fanfare, with all of the A-list hoteliers and visitors in attendance.[88] The *Gazette* reported:

> Built within two hundred yards of the historic [Luxor] Temple, its long wings facing close upon the river with its magnificent view, this latest addition to

> M. Pagnon's palaces in Upper Egypt is perhaps the finest and most elaborately-schemed hotel within the land of Egypt, and will accommodate some 200 guests. Facing the river on the west the whole of the back of the building looks over the garden and fertile ground to the mountains on the East, so that throughout the day the brilliant health-giving sunshine is pouring on to the balconies and terraces in the either one direction or the other. A huge out-stretched horseshoe terrace built on colonnades makes a welcome lounge overlooking where Nile steamers have their moorings, and double marble staircases lead to the entrance-hall.[89]

If the patio and gardens enticed visitors, the inside was even more luxurious. There was a bar and billiard room, wine cellars, dining and tea rooms, a music room, lounges, a restaurant, and a large back terrace overlooking the luscious green gardens that, at that time, grew fruit and vegetables for hotel and steamer patrons. For every three-bedroom-type room, there was a 'most complete bath and toilette room', as well as spacious apartment suites for wealthier guests who would stay the season.[90] There were also bedrooms and apartments for ladies' maids and valets of the wealthier patrons, and the domestic workers had their own dining room.

Even though it was an immediate hit with visitors, for short-term and long-term stays as well as for quick lunches, teas, meetings, and respites on hot days, Baedeker's *Egypt* did not update their map of Luxor in the 1908 edition to include the Winter Palace. The hotel is included in the list for Luxor that year, described this way: 'on the quay, to the S. of the landing-place, with lifts, terrace, and beautiful view towards the Nile',[91] but there is no outline of the building or even a dot marking its place on the map. It was not far from the Luxor Hotel and both had views of the ancient Luxor Temple. The 1914 edition of Baedeker's *Egypt*, the next in the series, shows the Winter Palace on the map, having taken over the spot on the page where the legend used to be (Figure 4.5). The hotel itself and gardens were shown to be huge – and this was not unrealistic. Many visitors would comment on marvelling at the building on the banks of the Nile as their boats or trains entered town.

Not all travellers appreciated the approach into Luxor once the Winter Palace took over the view, however. Travel writer and well-known curmudgeon Pierre Loti wrote this about his approach by dahabeah in 1910, just three years after the hotel was built:

> The thing which dominates the whole town, and may be seen five or six miles away, is the Winter Palace, a hasty modern production which has grown on the border of the Nile during the past year: a colossal hotel, obviously sham, made of plaster and mud, on a framework of iron ... painted a dirty yellow. One such thing, it will readily be understood, is sufficient to disfigure pitiably the whole of the surroundings.[92]

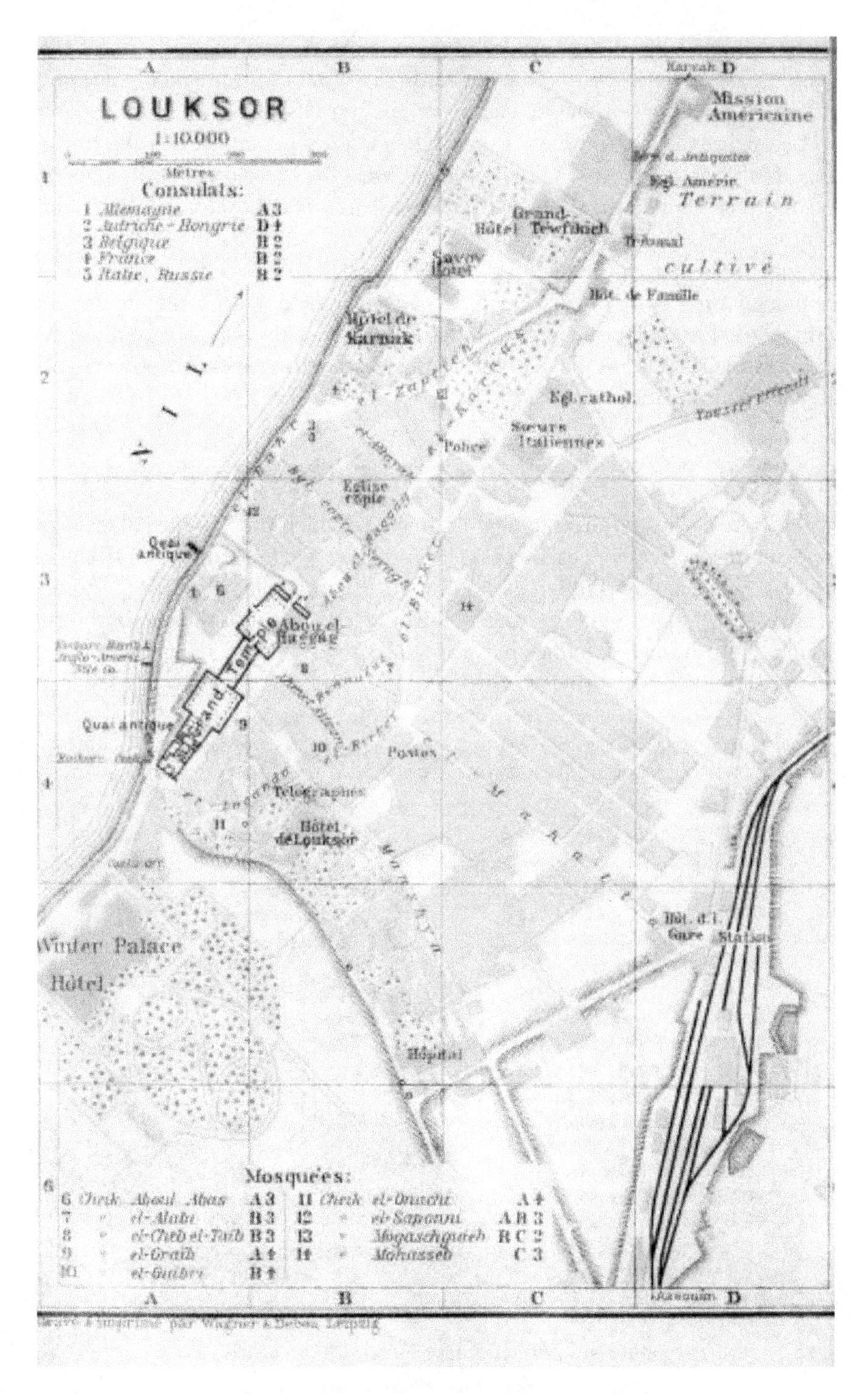

Figure 4.5 Baedeker’s map of Luxor 1914.

Despite what Loti or any other critics thought about the aesthetics of the building, as more tourists came to Luxor the Winter Palace quickly became a central place for archaeologists as well as tourists and journalists staying in Luxor for the season. Even for archaeologists who stayed in a dig house in Luxor on the West Bank, or who stayed on their own boat moored on the West Bank, the Winter Palace was, like Shepheard's in Cairo, the centre of Luxor Western social life; it was the place to see and be seen (Figure 4.6). For archaeologists, the Winter Palace quickly became and remained the centre of scientific networks of field work and an important place for them to meet. It became the stage on which a number of archaeological dramas played out between and among Western powers and Egyptian authorities.

Kings Valley Tomb 55

One of the earliest archaeological dramas for which the Winter Palace was a recurring character was the 1907 discovery by the Andrews and Davis crew of what is now known as KV 55. At the time, Davis swore the tomb was that of 'Queen Tyi [sic], and no mistake', as many of the tomb objects found there bore her name, her parents' names, or the names of others

Figure 4.6 Winter Palace Hotel, Luxor, c. 1920.

associated with Amarna royalty.[93] Current research suggests that the tomb was a cache of burial equipment from the Amarna royal necropolis and it is almost certain that the mummy in it was that of Akhenaten.[94] But Davis' botched excavation work has made identifying the tomb problematic, to say the least. Archaeologists today argue that the tomb almost definitely 'contained evidence to answer many of the persistent questions about the Amarna interlude', but the work Davis and Ayrton did on it was 'a disaster'.[95] Many scholars agree that the records of all the men digging that day 'left almost everything to be desired', thus it has transpired that Andrews' diary is the most complete narrative we have of the events at the excavation site.[96] Many times, as we will see, it is only because of Andrews' records that we know as much as we do about the scientific information of many of the sites they worked.

In the site report, based on Davis' memory and on Andrews' detailed diary of the events, Davis recalled: 'On the 1st of January, 1907, having exhausted the surrounding sites, I had to face a space of about forty feet oblong and at least fifty feet high, covered with limestone chippings, evidently the dumping of the surrounding tombs.'[97] These surrounding tombs were the open tombs of Ramesses IX, Seti I, and the Ramesses I, II, and III.[98] Even though he argued that 'it seemed to be a hopeless excavation', after almost a week of digging and one false start, their Department of Antiquities-assigned archaeologist, Edward Ayrton – accompanied by the Inspector of Upper Egypt at the time, Arthur Weigall – found the entrance to a tomb. Their photographer and artist Joseph Lindon Smith was there, as usual with their excavations, and helped to clear and guard the tomb opening throughout the process. The tomb entry and corridor were cleared in one day, and the following day, 9 January, the inner tomb entrance was breeched.

Due to the rapid and drastic change in climate combined with the fragility of the items in the tomb, as they cleared the entrance corridor and the tomb itself, they watched as many of the objects fell apart before their very eyes. Andrews wrote that the tomb was found 'in a state of great confusion', and that items were strewn about, but they were in 'quite good condition', covered with gold foil and easily read hieroglyphs.[99] Not only did she record who visited the site, who met on the dahabeah afterwards, as well as some of the items that were found, she also included her important personal reflections. David Erskine, an English MP, who had joined them by invitation, Weigall's wife Hortense, and Lindon Smith with his wife were all there for the opening of the tomb. Nettie Buttles was the first woman in the tomb, on Davis' orders. Moreover, Andrews detailed just how quickly the artefacts and the tomb itself were disintegrating. After a few more days of clearing and organising the artefacts in the tomb, Andrews recorded in her diary:

I went down to the burial chamber and it is now almost easy of access – and saw the poor Queen as she lies now just a bit outside her magnificent coffin, with the vulture crown on her head – all the woodwork of the shrine, doors, etc. are heavily overlaid with gold foil – which under the influence of the outer air is now peeling off – and I seemed to be walking on gold – and even the Arab working inside had some of it sticking in his wooly hair.[100]

Despite how much frustration this situation causes the present-day reader, Andrews' description is key in demonstrating a few things about this find. First, the tomb was exposed to the open air and quickly deteriorated. Second, no attempt was made to conserve or preserve artefacts and scholars have therefore lost a lot of important information, not to mention the gold foil everyone was walking on, and some walked away with in their hair. Third, with what they knew was such an important find, Andrews reported that a photographer was sent from Cairo (obviously not Joseph Lindon Smith) to record the tomb and there were dozens of pictures taken, but only seven appeared in the final printed report.[101] Archaeologists at the time criticised the parading through of visitors while the fragile artefacts were falling apart. Importantly, as Davis's biographer points out, Andrews

took the time to produce something none of the [other] archaeologists seem to have bothered with and Davis failed to include in his publication: a floor plan of the tomb mapping where the major artifacts had been located. Rough as it is, Emma's plan has been of value to Egyptologists ever since.[102]

Even though it was accurate and clear, Davis did not include her map in the excavation report that came the following year; neither Maspero, anatomist Grafton Elliot Smith, Ayrton, nor Davis mentioned Andrews or any of her contributions to this or any of the other excavations (Figure 4.7).[103]

During the course of their work, the group would go to the Winter Palace to relax with friends and possibly have a meal. They visited the hotel for the first time on 24 January, to have tea.[104] Andrews recorded that, both at the hotel and while working on the dahabeah, a lot of their discussion was about how to treat and preserve the artefacts coming out of the tomb, how to store them, and how to manage the Egyptian work crew.[105] The days following the tomb discovery were full of work at the site, attempting to save what could be saved. On 26 January, Andrews made her last trip down into the tomb, remarking that '[i]t was our last opportunity to see the tomb and the Queen. She is nothing but a mass of black dust and bones'.[106] Jones, newly full time on the crew, was finishing the drawings of the big door he had told his parents about before and Lindon Smith had finished his photographs.

On 29 January, Archibald Sayce joined Andrews on board the *Bedauin* for tea while Davis had taken some tourists to Medinet Habu. While there,

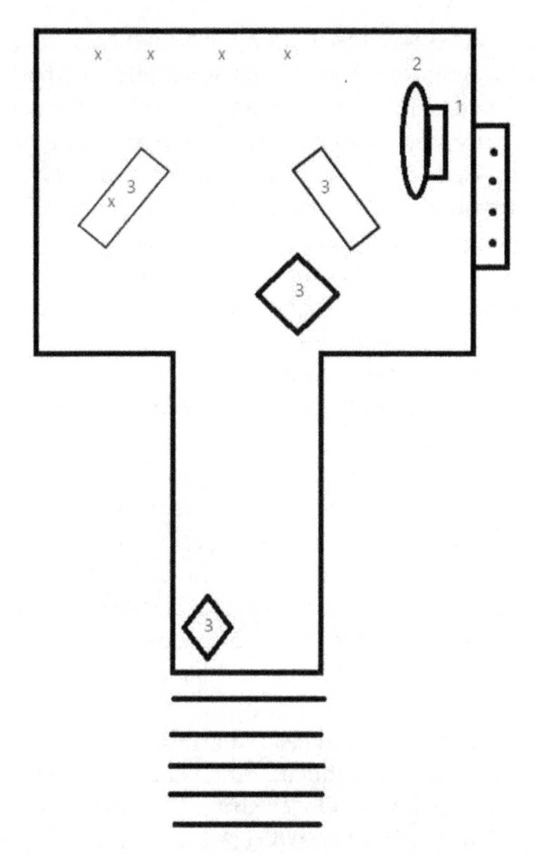

Figure 4.7 Adaptation of Emma Andrews' map of KV 55, Andrews' Diary, 19 January 1907.
Andrews noted rubbish strewn about the floor; labels include: (1) 'mummy of Queen', partly underneath the coffin (2); (3) large doors or panels with gold overlay, 'the one with the x has a beautiful portrait of Queen'. The Xs near the back wall are large panels, each with inscriptions and overlaid with gold. The niche to the right contained the four canopic jars.

he 'looked over many of the interesting things we have on board'.[107] The interesting things included the 'vulture crown', a crown of 'solid gold, and represents the royal vulture, with out-spread wings and meeting behind the head, beautifully done in a fine répoussé style – every feather perfect'.[108] The crown had been found when they opened the coffin to find the body 'wrapped in thick gold plates rather than foil'.[109] Andrews recorded they stored the crown 'in the closet at the head of my bed!'.[110] In fact, like Benson did at the Luxor Hotel, Davis and Andrews were storing a number of artefacts on board. For safety's sake they had a carpenter build them a large box, about eight and a half feet long by five feet long and high. Andrews remarked that the box and its contents were safe on their boat 'against everything but a concerted raid'.[111] Maspero arrived the following day to see the treasures, and then they had a veritable parade of friends and friends of friends on the *Bedauin* to see the items over the next few days. The Masperos and Ayrton were the most regular visitors, but Alfred Wiedemann and his wife and Frank Tarbell from Chicago stopped by to view the artefacts. Andrews, Davis, and the other Egyptologists and archaeologists shared information about the tomb, insight into their theories of the tomb, and more with each other and their visitors. By having these conversations away from the site in private and semi-private locations imbued with colonial power and privilege, they were excluding from the intellectual discussion the members of the crew who had done the physical work and who had critical local knowledge useful to the theories. In her diary, Andrews did not name the crew members who cleared passageways or who found any particular items, unless it was Davis, Maspero, or any other European member of their crew. Very deliberately, she did not name the man who walked away with the gold in his hair, but her description of his hair as 'woolly' would have informed readers that the person in question was likely from Sudan, and not Egypt, and may not have been important enough to her to name.

While most of the activity and cleared artefacts for KV 55 were centred on the *Bedauin*, they would frequently dine and meet friends at either the Luxor or Winter Palace hotels. On 2 February, Andrews went to the Luxor Hotel for tea and met 'Miss Hazard, President of Wellesley College, and Miss Bates Prof. of Literature of Wellesley. On our way home, we stopped at the Winter Palace, to see Miss Tuckerman … Harold Jones dined with us.'[112] Throughout the season, Andrews and Davis continued to have people over for breakfast, tea, and dinner. Andrews wrote that all of their visitors seemed 'perfectly enraptured with all the Egyptian talk and things'.[113] Jones remained working on their dahabeah, while going back and forth between there, the Valley, and the Luxor.

It is important to note here that this section has really focused on the excavation site and Andrews' and Davis' moored dahabeah, the *Bedauin*, as a meeting place for archaeologists. As I argued in the previous chapter, these private boats were crucial spaces for scientific discussions, training new crew, and displaying the work they had done that season. Andrews and Davis and their friends would continue to meet on the boat throughout their years in Egypt, but when the group returned to Luxor in December of 1907, the social scene had truly made its way to the Winter Palace, and so did the excitement.

'The Valley of the Tombs is now exhausted'

During the 1907–08 season, the team made some truly fascinating finds. They also had a dig house built out in the Valley, which was finished by December, and their archaeologist, Edward Ayrton, was living in one of its rooms. Despite Andrews thinking it was 'too damp to live in' at the time, the location was convenient for Ayrton; it saved time and money, and he got to work.[114] In late December, Ayrton found a shaft tomb, about thirty feet long near the entrance of the tomb of Seti I. It was full of gravel and Davis remarked that 'it was curious to see the decorative manner in which the mud had dried'.[115] Based on some of the items the group had found in this area over the past few years bearing Tutankhamun's name, Davis thought that the mud-packed tomb had been used previously by Tutankhamun, but later robbed and then used by Horemheb.[116] Andrews only remarked that 'it proved only a pit with large vases – some sealed – some more or less broken'.[117] The tomb was not really a tomb, even though it was given an official KV number (54). It turned out to be a collection of items used in the burial of Tutankhamun, now called the Tutankhamun embalming cache.[118] These items included a 'broken box containing several pieces of gold leaf stamped' with the names of Tutankhamun and his wife, Ankhsunamun on them, a blue cup with Tutankhamun's cartouche, an alabaster statuette, and a cloth with Tutankhamun's name.[119] There were pots and linens and bags of natron as well. Davis thought most of the items he found in the pit were too boring to care about. In fact, as Herbert Winlock later recalled, Davis was persuaded by Harold Jones to let the then-struggling MMA in New York have the lot.[120] A young Winlock packed up the items and sent them to New York; only three decades later was he able to study them.[121] The following season of 1908–09, around the same area, Jones found what Andrews recorded was 'a most lovely alabaster statuette of a woman – about 9 in. high, the most perfect specimen of Egyptian art I have ever seen, and in perfect condition. To think of her lying under that hard mud for nearly 3000 years!'.[122] Davis assumed he had found Tutankhamun's tomb, such as it was, and stopped searching for him after that.[123]

Another exciting development in the 1907–08 season happened on 3 January 1908. The team, led by Ayrton, found the 'mouth of a vertical shaft' near the tomb of Ramesses VI.[124] This tomb – a large, open room 'of a curious shape' with a small doorway at one side – was also filled with debris, so Davis and Ayrton used knives to chip away at the deposited mud.[125] They did not think much would come of this shaft, considering the seemingly lacklustre find they had just made a few weeks earlier, but about a week after discovering the shaft, on 12 January, Ayrton spotted glittering gold.[126] After using water to dissolve the surrounding packed mud, he found two gold earrings, marked with the name Seti II. They continued flooding the space to dissolve the mud, and Andrews wrote that they ended up finding a 'deposit of silver and gold objects of most unique importance and value ... containing cartouches of Sety II and Tausert – all of which Theo brought home [to the *Bedauin*] with him'.[127] From this site, George Daressy catalogued over 151 gold beads from a necklace and almost 80 other pieces of jewelry.[128] Harold Jones painted a number of these pieces and they are recorded in one of Davis' rare site reports, based, as usual, on Andrews' notes in her diary.[129] The tomb, now known as the Gold Tomb (KV 56) was ultimately seen, in itself, as dull because of its generally plain decoration. But it was truly unique in the gold and silver objects that were found throughout, the like of which would not be found until the tomb of Tutankhamun was uncovered fourteen years later.[130] The Gold Tomb is now believed to be the tomb of a child of Seti II and Tawosret.[131]

After the excitement of January, on 25 February 1908, Ayrton, whose work was sometimes hampered by the stubborn Davis, found the steps down to the entrance of the tomb of Horemheb.[132] Andrews recorded that Davis 'insisted upon entering the tomb, and going first. They found it very difficult and hot. It soon declared itself as the tomb of Horemheb.'[133] Andrews described it as 'a very long tomb – with corridors, well, chambers, etc. – walls beautifully ornamented with paintings, wonderfully fresh, of Gods, a magnificent sarcophagus, and figures in wood and marble – a few bones – no mummy – 200 ft long.'[134] Davis described it much the same (presumably because he got his notes from Andrews' diary), and wrote the sarcophagus as being 'made of red granite – 8 feet 11 inches in length, 3 feet 9 inches wide, and 4 feet in height – in perfect condition, and one of the most beautiful ever found'.[135] This tomb was designated as KV 57, and became the last tomb Davis and Andrews would find.

During all of this excitement in one of the pair's most productive seasons, they remained mostly on their dahabeah to entertain friends, show off their excavation artefacts, and have their crew work on the items and store them. They spent some time at the Winter Palace, dining and meeting friends and dignitaries who came to town. While Davis and Andrews frequently met with invited visitors on their boat, they also met a few people at the Winter Palace. They could control who was allowed on their boat and in their space, whereas hotels, being public yet colonial institutions, decided that for them. For example, although they invited the Luxor-based antiquities dealer Mohammed Mohassib for tea and other meetings on the *Bedauin*, and he had them over for tea and shopping in his store, Andrews and Davis never met him at the Winter Palace.[136] This was because Egyptians were not allowed on the terrace, in the gardens, or within the walls of the Winter

Palace unless they worked there or maintained considerable power within the antiquities department or other government ministry. Mohassib, despite his Western connections and importance in Egyptology (see Chapter 3, above), was still not included in the usual discussions of artefacts. On the other hand, they met their friends the Lewis M. Iddings family at the Winter Palace frequently. Iddings was the American Consul General to Egypt from 1905 to 1910, whose posting was in Cairo, but there was also a consulate in Luxor at the time. Davis frequently gave the family private tours of Karnak and the Valley, they were invited to dine on the *Bedauin*, and Andrews visited them at the hotel. On Boxing Day 1907, Andrews and their visitor that year, Carrie, went to the hotel to meet the Iddingses for lunch. Andrews remarked that the consular family 'had such pleasant rooms – and the whole house looked very attractive'.[137] They likely discussed the excitement happening in the Valley and how to deal with the pieces they had found in terms of American and Egyptian law.

After uncovering the tomb of Horemheb in February, the final one of the year, some of their friends came over to their boat, moored on the West Bank, but Andrews did not name who they were. It is possible it was the Iddingses, because, as Andrews wrote: '[t]hey said they had been watching from the [Winter Palace] Hotel opposite the royal entourage on the sand, waiting until it was over'.[138] It was possible for tourists to see the goings on of the *Bedauin* from the Winter Palace, as well as the boat's arrival and departure, centralising the hotel space as a truth spot from which to observe the parade of artefacts. It was also possible to watch the parade of people. As most people in Luxor knew who was arriving, and when, Harold Jones frequently waited for Andrews and Davis at the Winter Palace. When he saw them arrive, he went to their boat at once.[139]

We rely a great deal on Andrews' records of many of these finds and their activities while moored in Luxor, as well as during their trips on the dahabeah. Andrews worked behind the scenes, as it were, rarely if ever in photographs or dispatches from the field. There are currently no known extant photographs of Andrews.[140] However, from a passport application from her first trip to Egypt in 1887, we know that she was five feet, one inch tall, with brown hair and eyes, a fair complexion and an oval face with high eyebrows, and an aquiline nose over a medium mouth.[141] She sat in the shade, observing and writing – instead of digging, barging into tombs, and fainting in the heat like Davis seemed to enjoy doing. Her work was not acknowledged in any of the site reports that Davis produced. However, she was central to the success of Davis and their team. Not only did she organise and maintain their correspondence for all the years they dug in the Valley – as well as pay fully half the cost for at least two seasons of field work – but she was also the one who kept the all-important field journals that Davis failed to keep. These records are important to historians today for a number of reasons, not the least of which is that, for better or for worse, the dilettante millionaire Davis was very active in the excavations he funded and he worked long hours. It is because of her diaries that we even know about Andrews and the enormous role she played in Egyptology during this period.[142] Much of what Davis' biographer wrote about his Egyptology career comes from Andrews' diaries.[143] As important as her records are, they become sparse as the pair continued to make their annual return to Egypt. Often, even during later eventful excavation seasons, her records simply outline the people who visited the boat, and who dined with them. An important record for sure, but in terms of what they talked about, the later years were not covered in as much detail.

Harold Jones died in 1911, just a few years after their most significant season in 1907–08. He died from tuberculosis – the illness whose symptoms he sought relief from in Egypt in 1904 – so it was a slow, painful death to watch, as Andrews recorded in her diaries. By 1910, despite all the good his hard work and the dry air had done for his physical wellbeing, Jones' health was failing so much his brother came with him during his final season. In December of that year, Andrews wrote that 'Harold Jones is so poorly, so weak, quite incapable of work, even painting. But his brother Cyril keeps a certain supervision of the men.'[144] A week later, in an attempt to lift his spirits, Andrews threw a tea at the Winter Palace for Jones and invited a few of his friends. He made it through the tea, but 'the poor fellow looked very tired when it was over'.[145] The next time she mentioned him was that she learned of his death on 12 March: 'Came to our landing at 11.30 – and found the mooring garden looking very pretty. We arrived the 10th day from Luxor. Found telegram telling of Harold Jones' peaceful end on Thursday, the 9th.'[146]

Jones died while working on a project for Andrews and Davis in the Valley of the Kings. He had been instrumental in their work over the previous five years, especially in the tombs of Horemheb, KV 55 (the Amarna cache), and the Gold Tomb. His drawings of many of the objects are the best record archaeologists and historians now have of these items, due to Davis' lax preservation practices. His brother, Cyril, was with him when he died. Howard Carter and Lord Carnarvon helped Cyril bury his brother in Luxor.[147]

By 1912, Davis, Andrews, and the crew left had a largely unproductive season. In 1912, Davis published his final site report about the 1908 season, *The Tombs of Harmhabi and Touatânkhamanou*.[148] In it, Davis proclaimed: 'I fear the Valley of the Tombs is now exhausted.'[149] Davis' failing health, both his and Andrews' advancing age (they were both 75 years old), and being tired of all the travel were factors in their leaving as well. Every year from 1889 until 1913, they had spent around four months each in the

United States and Europe, with the other four months in Egypt. Andrews was constantly telling Percy Newberry how she hated to leave Newport so soon after they had arrived back home.[150] In 1914, the pair officially gave up their monopolising concession in the Valley after not having found anything for several seasons. It seemed that an era of excavation in the Valley was now over.

King Tutankhamun at the Winter Palace Hotel

The same year Davis and Andrews relinquished their concession, the First World War came to Egypt. This did not mean that Egypt was completely devoid of tourists and Egyptologists but, much like hotels in Cairo and Alexandria, hotels in Luxor were either forced to close for lack of business or used as British military headquarters or hospitals for casualties. Cook's luxurious steamers were used as troop transports and the opulent Winter Palace was converted into a convalescent hospital.[151] After the war was over, and Egyptologists were back in greater numbers, the Winter Palace remained the 'home base for the all-nations corps of Egyptologists' for decades.[152] Despite his ominous pronouncement about the Valley being exhausted, it was not long until Davis' and Andrews' uncovering of the spectacular Gold Tomb (KV 56) was outdone by Howard Carter, working for Lord Carnarvon.[153] Carnarvon, who began travelling to Egypt to relieve lingering pain from an automobile accident in England, was not only a regular at the Continental in Cairo, he was a fixture at the Winter Palace as well. He also made sure that his archaeologist had comfortable rooms there, and a dig house in the Valley.

Carter had worked in Egypt since 1891 and in the Valley since around 1905, with some of that time spent with Davis and Andrews. Once he left the Andrews and Davis crew, he had become used to exploring the hard-to-reach places on the West Bank for years, because Davis had the most sought-after concession, that is to say, the entire Valley. But when Davis gave up his concession, Carter pushed his patron Carnarvon to ask for it. He was ready to get into all the nooks and crannies Davis left unexplored so he might find more tombs. In 1916, Carter had learned that a tomb had been found by locals around seventy metres up a nearly vertical cliff face near the Valley. By the end of 1918, there had been a few more tombs found nearby as well.[154] In 1917, Carter and Carnarvon's search for the tomb of Tutankhamun in the Valley had begun in earnest. Carter wrote in the first volume of *The Tomb of Tut.ankh.Amen*, that 'the only satisfactory thing to do was to dig systematically right down to bed-rock'.[155] His goal was to make sure that no stone (or pile of stones) had been left unturned or undug. He suggested to Carnarvon that

> we take as a starting-point the triangle of ground defined by the tombs of Rameses II, Mer.en.Ptah, and Rameses VI, the area in which we hoped the tomb of Tut.ankh.Amen might be situated. It was rather a desperate undertaking, the site being piled high with enormous heaps of thrown-out rubbish, but I had reason to believe that the ground beneath had never been touched, and a strong conviction that we should find a tomb there.[156]

They continued to clear a number of layers in this exact area, and found some workmen's huts – meaning there was likely a tomb nearby – and some ostraca, all of which Carter called 'interesting but not exciting'.[157] Despite his lack of excitement, they continued to clear this area for two more seasons. They abandoned it for a season to try another spot, but the new site produced nothing. They returned to this triangle of ground in 1922, in what was their planned final season if nothing were to come of the Valley. In Carter's words, '[t]hat brings us to the present season and the results that are known to everyone'.[158]

This story has been told countless times. Even Carter, writing in 1923, tried to 'tell the story of it all', but knew he might not do it justice.[159] I will not repeat in great detail here the story that Carter and many others have pieced together, but I will try to summarise it as we know it so far. On 4 November 1922, it was likely Ahmed Gerigar, a *rais* on Carter's crew, who discovered a step cut in the rock floor.[160] The workmen dug, faster and faster, clearing away debris, to get to the bottom of sixteen steep steps cut into the floor of the valley. Here they found a sealed doorway. Some of the plaster had fallen away, so Carter could see into the passage beyond – it was full of stones and rubble. On 6 November, Carter cabled Carnarvon in England: 'At last have made wonderful discovery in Valley; a magnificent tomb with seals intact; re-covered same for your arrival; congratulations.'[161] Carter waited eighteen long days for Carnarvon to arrive, so they could continue to clear the steps and doorway. On 26 November, that which Carter called 'the day of days, the most wonderful that I have ever lived through', the team were able to clear the passage of debris and see into the doorway at the end.[162] Carter made a tiny hole in the upper left hand corner to test the air quality and to see if the next room or passage was full of debris, or empty. He put a candle in first, and, famously, described what he saw:

> At first I could see nothing, the hot air escaping from the chamber causing the candle flame to flicker, but presently, as my eyes grew accustomed to the light, details of the room within emerged slowly from the mist, strange animals, statues, and gold – everywhere the glint of gold. For the moment – an eternity it must have seemed to the others standing by – I was struck dumb with amazement, and when Lord Carnarvon, unable to stand the suspense any longer, inquired anxiously, 'Can you see anything?' it was all I could do to get out the words, 'Yes, wonderful things.' Then widening the hole a little further, so that we both could see, we inserted an electric torch.[163]

This is a favourite story Egyptologists and historians love to tell. But what Carter saw were more than just 'things'. On these material remains of a once vibrant civilisation were inscribed particular and problematic contexts. There was the problem of British–Egyptian relations; the issues of Western men claiming 'discovery' of these artefacts even though Carter and Carnarvon did very little of the actual digging. There was, and still is, the main discussion of to whom belonged not just the items themselves but the knowledge surrounding them.[164] All of this leads to the central questions of who gets to participate, how, and why?

The answer is both simple and complex. As historian Elliott Colla states,

> the discovery of Tutankhamen – Tut – goes well beyond the science of Egyptology to reach the entire range of political and expressive cultures in Egypt, from architecture to literature, from nationalist politics in the new Parliament to how Egyptian elites negotiated issues of sovereignty under British occupation.[165]

It further involves access to the scientific spaces, the knowledge about the science, and the language of the discussion, among many other factors. In and around Luxor at this time, there happened to be a virtual dream team of archaeologists, artists, linguists, and other experts, most of whom came to Carter's aid. This group included Britons Percy Newberry, Arthur Callender, and Alan Gardiner, as well as the whole MMA Egyptology excavation team who had been working nearby – including Herbert Winlock, Albert Lythgoe, and photographer Harry Burton, who would become famous for his over 2,000 images of the find.[166] Finally, and one of the main characters of this particular part of the story, there was the American Egyptologist James Henry Breasted.

After the First World War, as discussed in Chapters 2 and 3 above, thanks to a generous grant from the John D. Rockefeller Foundation, Breasted had been in Egypt and the Near East on a scientific and intelligence reconnaissance mission from 1919 to 1920, the first to make a tour of the new Near East that the aftermath of the War had created.[167] In late November 1922, Breasted was continuing the Epigraphic Survey he had begun in 1905, so he was in Luxor the day Carter saw the 'wonderful things'.[168] With his son, Charles, they stopped briefly in Luxor that day – just an hour or so spent to pick up a few supplies and mail – and sailed on to Aswan.[169] Upon their arrival in Aswan, a few days up the Nile from Luxor, Breasted received a letter from Carnarvon dated 3 December 1922. It read:

> 2 days after opening the cache or tomb I learnt you had been through Luxor. I wish I had known for I might then have persuaded you to stop a day & see a marvellous sight. Still there is another sealed door to be opened & I hope I should then have the pleasure of seeing you there. I expect this will take place some time in Feb. ... We have closed up everything for the moment as there are a good many preparations to be made (iron door preservation etc

> etc) & then Carter will get to work. It is a quite extraordinary discovery nothing has ever been seen like it & it is remarkable not only for the vast volume of stuff but for the beauty of the articles we have already seen, what we shall find when we get into the King's actual tomb I don't know.[170]

He did not have to tell Breasted *which* king he was talking about – Breasted would have known who they were looking for. Apparently, the Breasted party had missed seeing the opening of the door to the first chamber by only a few hours. They hurried back to Luxor, and booked their rooms at the Winter Palace Hotel. On 18 December, Charles and James emerged from the Winter Palace confident in their mission that day, but also certain that they were not to tell anyone where they were going.[171] Carter gave them clear, if round-about, instructions to get to the tomb, as 'a prevention of a swarm of bees following'.[172] Charles later recalled that:

> we did exactly as he told us, casually crossed the river, mounted donkeys and rode to the great temples and ruins along the margin of the western desert. Presently we left our donkeys at the foot of the cliffs, and as if merely to get the view, ascended the old familiar trail. With no one following us or aware of our errand, we continued climbing to the crest of the great ridge, thence descended at once on the other side. At the entrance pit of the new find we were met by Howard Carter and his assistant Mr. Callender; and Mr. Harry Burton, expert photographer, and Mr. A.C. Mace, field archaeologist, respectively from the Metropolitan Museum's Expedition at Thebes and Lisht, the services of whose entire staff had been lent to Lord Carnarvon by the Museum.[173]

It was an all-hands situation, and everyone went to help.

From this point on, as tourists and journalists poured into Luxor and descended on the Valley, the Winter Palace, as one of the largest and best-known hotels in town, was the busiest, even if it was the most expensive. Despite the fact that Carter had a dig house, as did the Metropolitan Museum (Chicago House was not yet complete), the Winter Palace Hotel became 'the clearing house for most of the complications and difficulties which now began to overtake Carter and his discovery'.[174] Carter's dig house was a popular stop for tourists by this point, as it was well known that he stayed there. By 1929 it was prominent on the Baedeker's *Egypt* map of the West Bank of the Nile.[175] It would not have been private for Carter and his confidants, not to mention not secure enough to keep anything there. It was therefore in the Winter Palace that Carnarvon met with heads of Egyptian ministries to discuss the excavations. He brokered agreements about the work, finances, finds, and politics over tea on the Winter Palace terrace. Carter and the Breasteds found solace in their rooms at the hotel, safe from reporters who had 'habitually divided their time between The Valley and the terrace of the Winter Palace Hotel, hoping for some new rumor or inadvertently dropped crumb of news which could be expanded into a cable dispatch'.[176] The men could and did lock themselves away so they could talk, write, and make important decisions regarding the tomb in the privacy afforded by the walls of the Winter Palace. In their private suites in this imposing symbol of European power, they reclaimed tenuous control over the narrative that was quickly escaping their grasp. In doing so, they made the Winter Palace one of the main locations of the unfolding drama. It was from this Western, colonial base that Carter would attempt to control access to the news and fight for the tomb and its contents.

In early 1923, during the first season clearing the tomb, Carnarvon sold the world copyright on news and images coming out of the Valley to *The Times* in London for £5,000 and 75 percent of future news royalties.[177] *The Times* then sold its Tutankhamun coverage to newspapers around the world, including back to newspapers in Cairo and Luxor.[178] The money made from these sales went to support the excavation and clearing of the tomb.[179] Carter claimed that Carnarvon did this to 'avoid constant interruption, and consequent dangers of work', but most people in journalism were not happy about this, and neither was the fledgling Egyptian Government.[180] From the time Egypt had been granted some independence by the British Foreign Office in February of 1922, there were violent power struggles between the two nations. The Tutankhamun find, administered by a British aristocrat and his British archaeologist, from the Winter Palace Hotel, brought these issues into sharp relief.[181] For instance, Egypt had been allowed to reorganise a number of government ministries, but Britain retained control over most diplomatic matters. Within this context of colonial conflict, when news broke of Carnarvon's lucrative deal with *The Times*, the Egyptian newspaper *Al-Ahram* maintained that *The Times* was forcing Egyptians to get news about their own history from a British outlet.[182] The news arrangement also ensured that Carter could dictate all the information himself, which granted him, and no one else, control over the knowledge coming out of the excavation.

In addition to the problem of access to and ownership of information was the difficulty of physical possession of the artefacts in the tomb and of the tomb itself. The tomb took over three years to completely record, clean, and empty of artefacts and even longer to pack, conserve, and send to the Cairo Museum; a full report on the tomb has never been published. In that short period of time, the new, nationalist-leaning government tightened antiquities laws, making it more difficult for Western excavators to retain finds.[183] These changes were much to the chagrin of many excavators in Egypt at the time, especially Carter and Carnarvon, who had hoped to continue the traditional (and original) concession agreement.[184]

When they received the concession for the Valley in 1914, excavation agreements were vague in terms of the division of finds. Often, artefacts were equally divided between the excavator and the Department of Antiquities, which would traditionally take 'objects of capital importance' for Egypt and leave a number of good pieces for wealthy patrons to sell or keep in order to entice them to continue donating time and money. The definition of 'capital importance' was historically lax and included 'unique' objects, but, ultimately, the division was completed at the discretion of the French Head of Antiquities.[185] This was generally how permissions had worked since the time of Mariette. Usually, excavators could use the situation to their advantage by negotiating with the Department of Antiquities to be lenient in the division. According to the original concession agreement with Carnarvon, he would have been able to bring a number of objects back to Britain to sell, display, and donate, much as he, Davis, and others had done for years.[186] But, because the laws changed around the original agreement, the vagaries became liabilities for both parties.[187] Pierre Lacau, Head of Antiquities from 1914, claimed that *every* item from the tomb fell under the 'objects of capital importance' clause and therefore Lacau retained everything excavated from the tomb for Egypt.[188] He did this partly to quell Egyptian officials' rising frustrations and to satisfy those who made arguments about nationalist claims. Carter and Carnarvon also did not like the fact that the new Egyptian Government – not particularly enthusiastic about the continued British presence in Egypt – was quickly taking control of their archaeological practice.[189] Breasted, Newberry, and others had written or spoken to the Ministry in support of Carter's work, but the political situation in Egypt was growing more volatile.[190]

Carter and Carnarvon had been fighting, not just with the Egyptian Government but also with each other about who had the rights to the artefacts coming out of the tomb.[191] They had mended their partnership, for the most part, but Lord Carnarvon's death on 5 April 1923 at the Continental Hotel in Cairo from a septic blood infection further complicated the situation. For Carter, losing his friend was difficult personally; losing his concession holder and powerful wealthy patron was a major problem professionally.[192] In Carnarvon's absence, Lacau happily granted a new, although more constraining, concession to Carnarvon's wife, Lady Almina. Yet there were struggles over this contract, and negotiations lasted for over a month. Breasted and other Carter surrogates worked out a new agreement on Carter's behalf that tried to accommodate both Carter and Egypt, but in the end, no one was happy with it.

By January 1924, Carter had grown tired of demands on him and his time, as well as the loss of control he felt in this excavation. He was weary of the mountains of correspondence, the crowds of official people demanding to see the tomb, and other unwelcome guests he was forced to entertain. He felt that 'the government's increasingly unreasonable demands were rapidly bringing their work in The Valley to a virtual standstill'.[193] The only way he knew to deal with the situation was to frequently withdraw to the privacy and relative safety of the Winter Palace. While there, he often confided in his trusted friend, James Breasted. Working from inside the hotel, the Breasteds, with Carter's knowledge and unspoken assent, started to strive to regain some creative control. Right around this time a journalist by the name of George Waller Mecham suddenly appeared in Luxor, filing articles with the *Chicago Daily News* and the *Christian Science Monitor.*[194] These pieces portrayed an insider's view of issues relating to the excavation and the troubles Carter was having with the Department of Antiquities and the Egyptian Government. They were critical of Egyptian officials' behaviour and decisions, which ended up further complicating the situation. Mecham, in fact, turned out to be none other than Charles Breasted, who was spending his days working with his father and Carter on the excavation, and his nights 'between midnight and 4 in the morning' in his rooms at the Winter Palace writing the articles – much of the information likely dictated by Breasted and Carter.[195]

Working together privately at the hotel in an effort to control the knowledge coming out of Luxor, their pieces gave a certain priority to these US-based newspapers, which in turn violated the concession agreement as well as the agreement Carnarvon had made with *The Times*.[196] Doing subversive work inside the Winter Palace made the hotel a central truth spot for the stories surrounding the Tutankhamun find itself. Being in the same room with Carter and watching the drama unfold in real time made the Breasteds' rooms the epicentre of forming and modifying the complex and specific knowledge about the state of the excavations and how the administration of them was progressing. The environment of the hotel as a place of Western power lent its authority to the British and American men, whose very presence reinforced the hotel's own dominance on the scientific landscape, and their own account of the perceived injustices they felt at the hands of the Egyptian ministers.[197] At this point, however, no one knew who Mecham was, so no one could stop him.

Despite the checks that the Department of Antiquities had placed on Carter, he continued with his work. On 12 February 1924, the British–American excavation team prepared to lift the lid of the outer yellow quartzite sarcophagus. As the group ate lunch in their separate, 'laboratory' tomb, KV 15, Carter read his correspondence for the day, including one dispatch from the antiquities department informing him of the 'tomb program for the next two days'.[198] With a shrug and a sigh, he set that missive aside for the moment, and went to see what the coffin would hold. With an intricate pulley system and what Breasted called 'an ingenious scaffolding', Carter raised the lid, unrolled the linen shrouds covering the next coffin, and revealed the gold outer coffin.[199] The whole process took about an hour. Everyone left the tomb that day ready for a productive season. Carter, expecting to go back into the tomb the following morning, left the one-and-a-half-ton quartzite lid hanging from its system of pulleys.

Instead of going to work, however, the next day Charles Breasted recalled that Carter 'burst in' to the Breasteds' hotel rooms at the Winter Palace, carrying orders from the Minister of Public Works in Egypt, Morcos Bey Hanna, stating that Carter was not to 'admit into the tomb of Tutenkhamon the wives or families of the collaborating scientists, as he had planned to do' that day.[200] As if it were an attempt to excuse an overuse of power, against ladies no less, by the Egyptian police, Breasted also fumed that the 'Minister had underscored these instructions by despatching an additional force of police to the tomb, so that if any American or English ladies invited by Carter had appeared, they would have been forcibly prevented from entering'.[201] As had become abundantly clear, the dispute over the tomb became about what Riggs terms 'politics, profit motives and prestige'.[202] Arguably, the Ministry wished to protect the find for these reasons. Carter also knew from his time on Davis' digs that too many visitors to a delicate site could destroy the artefacts. He felt, however, that Lacau and Morcos Bey Hanna were trying to have him removed from the job, and, in the meantime, to keep him from doing his job effectively, or at all. Carter was furious, as were the Breasteds, Mace, Lythgoe, Newberry, Alan Gardiner, a reporter from *The Times*, and others who had, by this time, gathered in the Breasteds' crowded Winter Palace suite.[203] According to Charles, Carter was 'fuming and pacing nervously up and down' his room. Carter chain-smoked while he dictated over a dozen versions of an announcement to Breasted until, with the help of the others, he finally decided upon the following:

> Owing to the impossible restrictions and discourtesies on the part of the Public Works Department and its Antiquity Service all my collaborators in protest have refused to work any further upon the scientific investigations of the discovery of the tomb of Tutankhamen.
>
> I am therefore obliged to make known to the public that immediately after the Press view of the tomb this morning between 10 A.M and noon the tomb will be closed and no further work can be carried out.
>
> (signed) Howard Carter[204]

The announcement was posted only in English all over Luxor, in the news, and spread around town by word of mouth, but it was *first* posted on a bulletin board in the lobby of the Winter Palace. Carter knew that the Winter Palace was one of the main intellectual centres of his work, and he knew what the hotel meant to the Egyptians and to those who used it as their powerful centralised base from which to do their work. He had hoped that the journalists and tourists who milled about the lobby would see it and take up his cause. As promised, after the press viewing that day, Carter locked the tomb. His professional colleagues tended to take Carter's side, even if they thought he was, as Alan Gardiner wrote, a 'difficult man, and by no means tactful'.[205] They did so because they felt that Lacau's duty as the head of the Department of Antiquities was to Egyptologists and their work, and *not* to Egypt or the Egyptian Government's interests. Nine days later, Charles Breasted wrote that the Egyptian Government 'sponsored and abetted' an 'official breaking in' of the tomb, and 'despite superlative clumsiness and many bleeding fingers', cut through Carter's locks and doors and replaced them with their own.[206] Except for using it as a site of a Wafd national celebration upon the opening of the new Egyptian Parliament, the tomb sat virtually unvisited until the end of March 1924.[207]

Despite the fact that Zaghloul's government was out by November of 1924, Carter had underestimated the power of the new Wafdist government. Further, he had not understood the ways in which public opinion could harm his case. In Egypt, both the Arabic and British press demonstrated that public opinion about European-dominated archaeology had soured quickly and Carter would have a fight on his hands.[208] During that March, there was a long legal battle between the Egyptian Government and Carter, with James Breasted acting as Carter's mediator. Because Breasted had been there from the beginning, and his rooms at the Winter Palace were central to the crafting of Carter's announcement to close the tomb, he seemed like the perfect choice.[209] In the end, Breasted garnered a new agreement between Egypt's government and Carter so that Carter could finish excavating the tomb. Carter had to agree that he never had intended to, and in the future never would, make a claim 'against the Egyptian government or against anyone else to any of the objects found in the Tomb of Tutankhamen'.[210] In the end, Carter regained the right to finish the excavation after he had made a tour of the United States in the spring and winter of 1924.[211] There was a newly negotiated and lasting agreement, which gave some right to Lady Carnarvon to 'duplicates ... wherever such duplicates can be separated from whole without damage to science'.[212] Carter signed the concession on 14 January 1925 knowing that there would be no such duplicates, but the agreement was with a new government who was more conciliatory to Carter and the British.[213] On 25 January, work resumed almost exactly where it had left off, except Carter had more control over the activities in the tomb than he had before.[214]

In 1927, Carter wrote of the new deal,

> much valuable time was wasted over a controversy singularly remote from the calm spirit which should guide research ... it became easier to work under

> the new conditions, especially when the rights of both parties were meticulously defined, and original claims of very old standing were forfeited for the good cause.[215]

He may have been right. Some of the vagaries of the original agreement were to blame for the issues between him and the Egyptian Government. However, this was the first clear battle that the newly independent Egyptian Government was able to win in order to gain control of their ancient past as well as the future of their history. The public excitement caused by the find gave the Egypt solid ground for controlling the rights to the country's past and decisions relating to it, with Lacau's support. Where Carter, the Breasteds, and other Europeans and Americans hoped to maintain control of the artefacts and the knowledge flowing out of Egypt, the newly independent government saw an opportunity to change the intellectual *and* political landscape. The issues of independence, nationalism, and ownership 'became embedded into the fabric of Egyptology and archaeology in Egypt'.[216]

These issues of politics, profit, and prestige were centred at the Winter Palace in the case of the Tutankhamun tomb. The Winter Palace was not only where Carter found his colleagues to talk about the excavation activity and plan their work, but also where he found a stronghold of European authority in a rapidly collapsing system. He went there because the Winter Palace was, in Luxor, the symbol of Western imperialism, a place which allowed him both comfort and power. Carter, Breasted, and the other Westerners who worked on the tomb and throughout Egypt were navigating a whole new system in the post-War world. They did it in and from the comfort of their hotel rooms, in the domestic space where they maintained some control over the outcome of the policies they were manufacturing for implementation in public and professional places.

While it is difficult to say what Carter's memo or Charles Breasted's/ George Mecham's articles would have looked like if they had been written and posted first in a more public liminal space, such as Carter's public dig house, or a more professionally controlled space, such as a university or museum office, there are particular traits we are able to apply to the private hotel space. Because hotels in Egypt were sites of Western authority, power, and control, the policies, ideas, and actions that came out of them were imbued with the expectation that they would be held in high esteem. Carter expected that the construction of the Winter Palace as a truth spot in the colonial political system and the Egyptological world would bolster his claims over the tomb.

Carter and the Breasteds' private discussions combined with the participation of other Western men in the cognitive topography present in Luxor during this period offered Carter the moral and political support he had become accustomed to. One thing that surprised these men was that they did not know how to navigate the new power structure in Egypt. Carter and others were being supervised by Egyptians, the very people they had been instrumental in disenfranchising from their own history for over a century. Almost everyone involved in the situation on both sides knew that the fight over King Tutankhamun's tomb would set the precedent for all future decisions over excavation rights. That is why they fought so hard from their mighty Palace fortress to obtain rights to continue their work on the project, much less own anything from it.

Conclusions

This chapter is longer and more complex than the others, but it is because the power, role, and impact of hotels on the creation of the discipline is so evident in Luxor. There were more excavations and a lot more archaeological activity going on in the vicinity of the Luxor Hotel and Winter Palace, so the documentary evidence is more prevalent in this area. In the story of Egyptology, hotels were central places where knowledge about a find was mediated, tested, discussed, transformed, and from where it was spread.

In the case of the Temple of Mut at Karnak, the Luxor Hotel was a dig house and respite for the Bensons, their family, and friends. They stored their finds there as one would have done in a dig house. They met new crew members there, worked to train them, and wrote all of their correspondence from their rooms or the luxurious gardens. While the work of the excavation was obviously done on site, the Luxor Hotel was a central node and truth spot for this work, as well as the building of Benson's entire network of Egyptologists she used to work on the Temple. She met and fell in love with Janet Gourlay at the Luxor Hotel, and her family stayed there with her when they came to visit after her father's death. This personal activity made the hotel an important domestic space for Benson. It was a comfortable space for meeting her friends and family, as well as new love.

Davis and Andrews used both the Luxor Hotel and the Winter Palace for a number of discipline-building activities. They met with, wooed, and hired Harold Jones and Howard Carter in the confines of the gardens at the Luxor Hotel. It is interesting to think that they must have run into Benson and her crew when they were there. Maybe they passed each other in the lobby, or sat next to them on the terrace during tea. As we have no evidence of this, we cannot do more than speculate. But we do know that they used the hotel for similar purposes. The Winter Palace was clearly a colonial space in which ideas were discussed, formed, and from which they were disseminated. The most obvious example being that of the information and control of that information surrounding the discovery of Tutankhamun's tomb.

Since 1922, Egypt has maintained control over archaeological activities in the country.[217] The Department of Antiquities have begun to ask for many of the pieces that were taken during the direct colonial control of Egyptology to be returned.[218] That conversation is still ongoing. In the midst of the frenzy that arose around Carter, Tutankhamun, the gold, and the news, one central calm place of respite, reflection, and work was the Winter Palace. From rooms in the hotel, Carter worked with the Carnarvons, the Breasteds, the government, and others to try to control the work and the information that flowed through the hotel. Tutankhamun himself became a symbol of nationalism in Egypt – the power and glory of Egypt's past that can carry on into its present and future. The Winter Palace still stands in Luxor—its façade, ageing and in need of some repairs, as imposing as ever on the Nile.

Notes

1 Baedeker's *Egypt* (1892), 101.
2 *Ibid.*
3 Edwards, *A thousand miles up the Nile*, 176–7.
4 Andrews' Diary, 11 January 1890.
5 Baedeker's *Egypt* (1908), 247.
6 *Ibid.*, 204.
7 Carruthers, 'Credibility, civility, and the archaeological dig house'; Morgan and Eddisford, 'Dig houses'.
8 Blake C. Scott, 'Revolution at the hotel: Panama and luxury travel in the age of decolonisation', *Journal of Tourism History* 10:2 (2018): 149.
9 For crucial work being done on the exclusion of Egyptian excavators from the European histories of Egyptology, see, for example, Heba Abd el-Gawad and Alice Stevenson, 'Egypt's dispersed heritage: Multi-directional storytelling through comic art', *Journal of Social Archaeology* (20 February 2021). https://doi.org/10.1177%2F1469605321992929. Also, Rachel Mairs and Maya Muratov, *Archaeologists, tourists, interpreters: Exploring Egypt and the Near East in the late 19th–early 20th centuries* (London: Bloomsbury, 2015).
10 Thomas Cook, Ltd., *Thomas Cook & Son in Egypt, 1869–1926* (Thos. Cook, 1926), 3. Black Box 1, Cook archives.
11 Cook, *Thomas Cook & Son in Egypt*, 7. Black Box 1, Cook archives.
12 Brendon, *Thomas Cook*, esp. 120–40.
13 Baedeker's *Egypt* (1892).
14 See Hinrichsen, *Baedeker's Travel Guides*, 46–8.
15 Baedeker's *Egypt* (1898); Hinrichsen, *Baedeker's Travel Guides*, 47.

16 For example, *The Egyptian Gazette* reported on 8 March 1900 that Maspero and Sayce, among others, were arriving at Luxor and staying at the Luxor Hotel.
17 Loti, *Egypt*, 157, 181.
18 *Ibid.*
19 Humphreys, *On the Nile*, 61.
20 Humphreys, *Grand hotels*, 176.
21 *Ibid.*
22 *Ibid.*
23 John M. Cook to Bert Cook, 10 January 1889, Black Box 8, Cook archives.
24 *Ibid.*
25 *Ibid.* It's unclear what the sign said, because it is not included in the letter.
26 Fran Weatherhead, 'Painted pavements in the Great Palace at Amarna', *The Journal of Egyptian Archaeology* 78 (1992): 179.
27 John M. Cook to Bert Cook, 10 January 1889, Black Box 8, Cook archives.
28 *Ibid.*
29 *Ibid.*
30 *Ibid.*
31 Andrews' Diary, 11 January 1890.
32 *Ibid.*
33 Baedeker's *Egypt* (1892), 101–2.
34 Sylvie Weens, 'Tales of antiquities at the Luxor Hotel', *Ancient Egypt* (February/March 2019): 16–21.
35 Baedeker's *Egypt* (1892), 101–2.
36 *Ibid.*
37 Margaret Benson and Janet Gourlay, *The Temple of Mut in Asher* (London: John Murray, 1899); Benson, ed. *Life and letters*.
38 Maggie Benson to her Mother, 11 March 1894, in *Life and letters*, 177.
39 *Ibid.*
40 *Ibid.*, 417–18.
41 Auguste Mariette, *Karnak: étude topographique et archéologique avec un appendice comprenant les principaux textes hiéroglyphiques découverts ou recueillis pendant les fouilles exécutées à Karnak* (Leipzig: J.C. Hinrichs, 1875).
42 Maggie Benson to her mother, 2 January 1895, *Life and letters*, 190.
43 *Ibid.*
44 Benson and Gourlay, *Temple of Mut*, 26.
45 Maggie Benson to her mother, 2 January 1895, *Life and letters*, 190.
46 Benson and Gourlay, *Temple of Mut*, 32.
47 *Ibid.*, 32–3, 189–90.
48 Maggie Benson to her father, 13 February 1895, *Life and letters*, 192–3.
49 Benson and Gourlay, *Temple of Mut*, 35. William Peck notes that the statue's number at the Egyptian Museum in Cairo is Cairo CG No. 566.
50 Lady Jane Lindsay to Benson's mother, 1 February 1896, *Life and letters*, 198.
51 Sheppard '"Constant Companions" and "Intimate Friends."' See also Simon Goldhill, *A very queer family indeed: Sex, religion, and the Bensons in Victorian*

Britain (Chicago: University of Chicago Press, 2016); Sharon Marcus, *Between women: Friendship, desire, and marriage in Victorian England* (Princeton: Princeton University Press, 2007); Martha Vicinus, *Independent women: Work and community for single women, 1850–1920* (Chicago: University of Chicago Press, 1985); Martha Vicinus, *Intimate friends: Women who loved women, 1778–1928* (Chicago: University of Chicago Press, 2004).

52 Maggie Benson to her Mother, 18 May 1896, *Life and letters*, 206–7.
53 Benson and Gourlay, *Temple of Mut*, 38–9.
54 Maggie Benson to her mother, 14 February 1896, *Life and letters*, 202.
55 Benson and Gourlay, *Temple of Mut*, 59.
56 http://williamhpeck.org/e_f_benson_in_egypt (accessed 3 April 2020).
57 Benson and Gourlay, *Temple of Mut*, 64.
58 *Ibid.*, 65, 357.
59 *Ibid.*, 27, 36.
60 Baedeker's *Egypt* (1898), 225. http://williamhpeck.org/e_f_benson_in_egypt (accessed 3 April 2020).
61 Maggie Benson to Nettie Gourlay, 10 August 1896, *Life and letters*, 216.
62 Maggie Benson to Nettie Gourlay, 28 August 1896, *Life and letters*, 220.
63 Vicinus, *Intimate friends*, 134–41.
64 See Elizabeth Waraska, *Female figurines from the Mut Precinct: Context and ritual function* (Göttingen, Vandenhoeck & Ruprecht, 2009); 'Brooklyn Museum: Temple of Mut', www.brooklynmuseum.org/features/mut (accessed 30 October 2020).
65 Anna Garnett, 'John Rankin and John Garstang: Funding Egyptology in a Pioneering Age', in *Forming material Egypt: Proceedings of the International Conference, London, 20-21 May, 2013*, eds P. Piacentini, C. Orsenigo, and S. Quirke (Milan: Pontremoli, 2013–14), 96.
66 Delany, *A son to Luxor's sand*, 5. See Chapter 2 for more information about Jones' coming to Egypt and his work with Garstang.
67 Harold Jones to Family, Sunday [8] January 1905. Garstang Museum Archives, FC/1/4 Xeroxes of Ernest Harold Jones. Chapter 3, above.
68 *Ibid.*
69 Harold Jones to Family, 14 January 1905. Garstang Museum Archives, FC/1/4 Xeroxes of Ernest Harold Jones.
70 Harold Jones to Family, 16 February 1905. Garstang Museum Archives, FC/1/4 Xeroxes of Ernest Harold Jones.
71 Harold Jones to Family, 2 March 1906. Garstang Museum Archives, FC/1/4 Xeroxes of Ernest Harold Jones.
72 *Ibid.* See the conversion Table 0.1: £15 per plate would be the equivalent of £1,248 or $1,727 per plate today.
73 See the conversion Table 0.1. This would be equivalent to £16,650, or $23,042 today. According to the National Archives purchasing power conversion tool, this was an amount equivalent to 606 days of work for a skilled tradesperson, which Jones arguably was.
74 Harold Jones to Family, 12 March 1906. Garstang Museum Archives, FC/1/4 Xeroxes of Ernest Harold Jones.

75 See Chapter 2.
76 Harold Jones to Family, 7 December 1906. Garstang Museum Archives, FC/1/4 Xeroxes of Ernest Harold Jones.
77 Harold Jones to Family, 7 December 1906. Garstang Museum Archives, FC/1/4 Xeroxes of Ernest Harold Jones.
78 Harold Jones to Family, 3 February 1907. Garstang Museum Archives, FC/1/4 Xeroxes of Ernest Harold Jones.
79 *Ibid.*
80 *Ibid.* Given Jones' penchant for misspelling names, this was most likely Ernest Gardner, as Gardiner was a student at Oxford from 1906–08 and had visited Egypt in 1908. Gardiner would not have had his own tent with an entourage of well-known Egyptologists on site.
81 Harold Jones to Family, 3 February 1907. Garstang Museum Archives, FC/1/4 Xeroxes of Ernest Harold Jones.
82 *Ibid.* Carter used rooms at the Karnak and later at the Winter Palace to sell his paintings between excavation jobs (Winstone, *Howard Carter*, 93).
83 Harold Jones to Family, 3 February 1907. Garstang Museum Archives, FC/1/4 Xeroxes of Ernest Harold Jones.
84 Andrews' Diary, 23 January 1907.
85 Andrews' Diary, 13 February 1907.
86 Andrews' Diary, 8 February 1907; see also Ketchley, 'Witnessing the Golden Age'.
87 Carter had recently survived a few scandals, including issues with French tourists in Saqqara. See Winstone, *Howard Carter*, 77–95.
88 Humphreys, *Grand hotels,* 180.
89 'Upper Egypt Hotels Co. Inauguration of Winter Palace', *The Egyptian Gazette* (19 January 1907).
90 Humphreys, *Grand hotels*, 180.
91 Baedeker's *Egypt* (1908), 248.
92 Loti, *Egypt*, 180.
93 The wife of Amenhotep III. Andrews' Diary, 9 January 1907.
94 Thompson, *Wonderful things, Vol.* 2, 254; C. Nicholas. Reeves, *The Valley of the Kings* (New York: Kegan Paul, 1990), 42–9; Martha R. Bell, 'An armchair excavation of KV 55', *Journal of the American Research Center in Egypt* 27 (1990): 97–137.
95 Thompson, *Wonderful things, Vol.* 2, 252.
96 *Ibid.* Bell used the typed version in the MMA as a main source in her reassessment of the tomb in 1990.
97 Theodore Davis, 'The finding of the tomb of Queen Tîyi', *The tomb of Queen Tîyi* (London: Duckworth, 1908), 1.
98 *Ibid.*
99 Andrews' Diary, 9 January 1907.
100 Andrews' Diary, 19 January 1907.
101 It is not clear where these pictures went, but they could show archaeologists the original state of many of the artefacts, making study of this period more productive.

102 Adams, *The millionaire and the mummies*, 146.

103 The report itself is problematic, as Davis was sure that the mummy was that of a woman and tried to get Elliot Smith to say as much. When the report came out, Elliot Smith argued that 'the skeleton is that of a man of twenty-five or twenty-six years of age' (Grafton Elliot Smith, 'A note on the estimate of the age attained by the person whose skeleton was found in the tomb', *The tomb of Queen Tîyi*, xxiv). Despite this argument, the report continued as if the tomb belonged to Queen Tiye. Elliot Smith had stated that he could make a more detailed argument for his case, but he did not do that here, as 'Mr. Davis tells me he does not want [that] for the purposes of this volume' (xxiv).

104 Andrews' Diary, 24 January 1907.

105 Andrews' Diary, 22–25 January 1907.

106 Andrews' Diary, 26 January 1907.

107 Andrews' Diary, 29 January 1907.

108 Andrews' Diary, 25 January 1907.

109 *Ibid.*

110 *Ibid.*

111 Andrews' Diary, 29 January 1907.

112 Andrews' Diary, 2 February 1907.

113 Andrews' Diary, 3 February 1907.

114 Andrews' Diary, 21 December 1907.

115 Theodore M. Davis, *The tombs of Harmhabi and Touatânkhamanou: The discovery of the tombs* (London: Constable and Co., 1912), 2.

116 *Ibid.*, 3.

117 Andrews' Diary, 21 December 1907.

118 Herbert E. Winlock, *Materials used at the embalming of King Tutankhamun*, Metropolitan Museum of Art Papers, No. 10 (New York: The Metropolitan Museum of Art, 1941).

119 Davis, *The tombs of Harmhabi and Touatânkhamanou.*

120 Winlock, *Materials used at the embalming of King Tutankhamun*, 6.

121 *Ibid.*

122 Andrews' Diary, 21 January 1909.

123 Davis, *The tombs of Harmhabi and Touatânkhamanou*, 3.

124 Davis, 'The unnamed Gold Tomb', in *The tomb of Siphtah,* 31.

125 *Ibid.*

126 Andrews' Diary, 12 January 1908.

127 *Ibid.*

128 George Daressy, 'Catalogue of the jewels and precious objects of Setuî II and Tauosrît found in the unnamed tomb', in *The tomb of Siphtah*, 33–46.

129 Davis, 'The unnamed Gold Tomb,' in *The tomb of Siphtah*, 31–46, and see plates.

130 Adams, *The millionaire and the mummies*, 202.

131 *Ibid.*

132 *Ibid.*, 206–7.

133 Andrews' Diary, 29 February 1908.

134 *Ibid.*
135 Davis, *Tombs of Harmhabi and Touatânkhamanou*, 2.
136 Andrews' Diary, 6 March 1908.
137 Andrews' Diary, 26 December 1907.
138 Andrews' Diary, 2 March 1908.
139 Andrews' Diary, 1 December 1908.
140 See Ketchley, 'Witnessing the Golden Age'. Ketchley has located an image with Andrews in it, but the photo is too grainy to see her face in the distance. There is another image that *may* be of Andrews, but Ketchley is not convinced.
141 Emma B. Andrews Diary Project: www.emmabandrews.org/project/ (accessed 19 November 2020).
142 At the time of writing, Andrews did not leave much in terms of a correspondence collection of her own.
143 Adams, *The millionaire and the mummies.*
144 Andrews' Diary, 11 December 1910.
145 Andrews' Diary, 18 December 1910.
146 Andrews' Diary, 12 March 1911.
147 Delany, *A son to Luxor's sand.*
148 Davis, *Tombs of Harmhabi and Touatânkhamanou.*
149 *Ibid.*, 3.
150 Emma Andrews to Percy Newberry, 22 May 1902, NEWB2/019, Griffith Institute of Egyptology, Oxford University.
151 Humphreys, *Grand hotels*, 183.
152 *Ibid.*
153 C. Breasted, *Pioneer to the past*, 327.
154 Nicholas Reeves and John H. Taylor, *Howard Carter: Before Tutankhamun* (New York: Harry Abrams, 1993), 129–31.
155 H. Carter and A. C. Mace, *The tomb of Tut-ankh-Amen, discovered by the late Earl of Carnarvon and Howard Carter, Vol. 1* (New York: George H. Doran, Co., 1923), 124.
156 *Ibid.*, 124–5.
157 *Ibid.*, 125.
158 *Ibid.*, 128. There is a lot that has been written about this episode in the history of Egyptology, so I will not cite it all. However, some important works include C. Breasted, *Pioneer to the past*, 327–49; more recent are Colla, *Conflicted antiquities*, 172–226; Reid, *Contesting antiquity*, 51–133; Riggs, *Photographing Tutankhamun.*
159 Carter and Mace, *The tomb of Tut-ankh-Amen, Vol. 1*, 131.
160 The story began, alternatively, with a water boy who stumbled on the step. Christina Riggs' discussion of the problems in this story point to the fact that it was likely *rais* Ahmed Gerigar who found the step and was in charge of much of the actual excavation of the stairs down to the entrance of the tomb (*Photographing Tutankhamun*, 9–10; 154–60).
161 Carter and Mace, *The tomb of Tut-ankh-Amen, Vol. 1*, 136.
162 *Ibid.*, 141.

163 *Ibid.*, 141–2.
164 See Colla, *Conflicted antiquities*, 172–226.
165 *Ibid.*, 177–8.
166 Christina Riggs's in-depth study of this must always be mentioned, *Photographing Tutankhamun*. She bases her study on the photographs in the Griffith Institute of Egyptology, Oxford University.
167 Sheppard, 'On His Majesty's Secret Service'.
168 C. Breasted, *Pioneer to the past*, 330. This was the day that Carter told Carnarvon he saw 'wonderful things'.
169 C. Breasted, *Pioneer to the past*, 330–4.
170 Lord Carnarvon to JHB, 3 December 1922, Box: DOC III WB4 – Box A 1922, A-R. JHB Correspondence.
171 C. Breasted, *Pioneer to the past*, 335.
172 Howard Carter to JHB, 16 December 1922, Box: DOC III WB4 – Box A 1922, A-R. JHB Correspondence.
173 *Ibid.*
174 *Ibid.*, 358.
175 Baedeker's *Egypt* (1929), after 270.
176 C. Breasted, *Pioneer to the past*, p. 358
177 See the conversion Table 0.1. In 1922, this amount would be the equivalent of around £154,000 or $213,000 today.
178 Reid, *Contesting antiquity*, 63–4; Riggs, *Photographing Tutankhamun*, 18–32.
179 C. Breasted, *Pioneer to the past*, 346.
180 H. Carter and A. C. Mace, *The tomb of Tut-ankh-Amen, Discovered by the late Earl of Carnarvon and Howard Carter, Vol.* 2 (New York: George H. Doran, Co., 1927), x; Jason Thompson, *Wonderful things: A history of Egyptology, Vol. 3: From 1914 to the twenty-first century* (Cairo: American University in Cairo Press, 2018), 55–6.
181 Reid, *Contesting antiquity*, 53–5.
182 *Ibid.*, 64.
183 *Ibid.*, 65–6.
184 Thompson, *Wonderful things, Vol. 3*, 62–4.
185 Goode, *Negotiating for the past*, 70–2.
186 Adams, *The millionaire and the mummies.*
187 Carter and Mace, *The tomb of Tut-ankh-Amen, Vol.* 2, x.
188 Riggs, *Photographing Tutankhamun*, 24.
189 C. Breasted, *Pioneer to the past*, 347.
190 Riggs, *Photographing Tutankhamun*, 25.
191 Winstone, *Howard Carter*, 194–7.
192 In September of 1925, Carter was given Carnarvon's old rooms when he stayed at the Continental. The manager gave the rooms to Carter in memory of his old friend (Humphreys, *Grand hotels*, 129).
193 C. Breasted, *Pioneer to the past*, 355.
194 For example, G. W. Mecham, 'Tut's Tomb becomes Mecca for Tourists', *The Chicago Daily News* (6 February 1924): 2.

195 C. Breasted, *Pioneer to the past*, 355.
196 Abt, *American Egyptologist*, 312.
197 See Reid, *Contesting antiquity*, 55–61 for a succinct depiction of the colonial views both Europeans and Americans held of Egyptian ability to govern themselves or run an excavation.
198 C. Breasted, *Pioneer to the past*, 361; Riggs, *Photographing Tutankhamun*, 21–2.
199 C. Breasted, *Pioneer to the past*, 363.
200 *Ibid.*, 365–6.
201 *Ibid.*
202 Riggs, *Photographing Tutankhamun*, 25.
203 James, *Howard Carter*, 337; Winstone, *Howard Carter*, 228–9.
204 C. Breasted, *Pioneer to the past*, 366.
205 James, *Howard Carter*, 343.
206 James, *Howard Carter*, 347; C. Breasted, *Pioneer to the past*, 367. Breasted continued to publish articles in the *Chicago Daily News* throughout the ordeal, all as Mecham, for example: G. W. Mecham, 'Breasted in Protest over Tut Tomb Ban' (14 February 1924): 2; G. W. Mecham, 'Tut Tomb Discoverer Barred Out By Police' (15 February 1924): 2.
207 Reid, *Contesting antiquity*, 71.
208 *Ibid.*, 68–74; Goode, *Negotiating for the past*, 84–5.
209 James, *Howard Carter*, 349.
210 Ibid.
211 Winstone, *Howard Carter*, 243–65.
212 Thompson, *Wonderful things, Vol. 3*, 70.
213 Riggs, *Photographing Tutankhamun*, 27.
214 Goode, *Negotiating for the past*, 90–1.
215 Carter and Mace, *Tut-ankh-Amen, Vol. 2*, xii.
216 Thompson, *Wonderful things, Vol. 3*, 70.
217 This is not to say that artefacts did not leave Egypt. See Stevenson, *Scattered finds*, 259–60.
218 For example, Tom Teicholz, 'The British Museum: The Problematic Yet Enduring Appeal of Antiquities', *Forbes* (14 October 2018).

Conclusion: going back home

Luxor is the last stop in this book. It is true that many tourists and Egyptologists travelled much further south throughout their time in Egypt. Like the Breasteds, the Bensons, Amelia Edwards, and Helen Tirard, they went to Aswan, to and through the First Cataract, and some on to Sudan. The sites further upriver include the temple island of Philae, Abu Simbel, and the pyramid fields of Meroë. But for the purposes of this book, much like for the tourists on Cook's luxury steamers, the bigger sites must do for now.

After their excavation seasons, Egyptologists would convene once again in Cairo to divest themselves of their crews, split artefacts with the museum, make final arrangements for their shipments going back, and make more plans for the following season. As we have seen, some of them stayed in Cairo as long as they could to avoid the English summer. Harold Jones and Howard Carter frequently stayed as long as they were able and the money and work held out. As Andrews, Davis, Wilbour, and Sayce unpacked their dahabeahs and their crews worked to get them ready for the next year, they visited with friends and finished souvenir shopping for their return home.

The travellers usually left the same way they came: through Alexandria, back across the Mediterranean to any of the ports that would welcome them back to Europe, Britain, and the United States. Americans like the Breasteds and Andrews and Davis slowly made their way home through European cities to meet with colleagues, family, and friends. They would often visit important museums and collections to continue their research. Often, because these were larger, more established institutions, not to mention the fact that they were on their home (Western) soil, these parts of their travel tended to be a bit more constrained by professional norms. They had left the field and the liminal spaces of the hotels and travel behind.

The book has been a historical geography and a phenomenological travelogue. I have argued that we must contextualise the practice of Egyptology not just in museums, classrooms, and on sites, but also as a fluid process of travel, professionalisation, and conversation. These processes were performed by Western people who were often new to Egypt, new to the discipline, and who needed connections to powerful people and networks to get their careers started, such as they were. By contextualising their activities in these places and spaces, we can analyse these familiar sites and people in a fresh light. The hotels, boats, and other meeting spots addressed here were central to the discipline as peaks on the topography of the scientific landscape and pivotal nodes of scientific networks in Egypt. They were particularly suited for these activities because of the Western colonial power wielded by the British in these spaces and how practitioners of archaeology in Egypt benefitted from that power.

I organised the book as though it were a travelogue, embarking up the Nile. As travellers arrived in Alexandria, the city itself was significant as an entry space. The historical tourism of this city, as well as the chapter, set up the travellers for all of the sites they would encounter along the way as both tourists and scholars. Over time the city grew, as did the amount of knowledge archaeologists understood about the history of the city. Growing knowledge impacted the amount of time travellers would then spend in the city – both increasing the permanent population as well as the transient tourist population who would crowd Alexandria's ports in October and April, coming and going. There is much more study due on the history of Alexandria within the history of Egyptology. Emprereur and Golvin have done beautiful work in this vein, and the video game *Assassin's Creed: Origins* presents a detailed and accurate view of the city in the Ptolemaic period.[1] But there are still so many more questions than answers.

Cairo was clearly a centre of power, not only for the colonial then nationalist government but also for the archaeologists and hopeful professionals who arrived there. It was a staging ground for Western tourists during this time period. They would prepare for their trips up the Nile for days to weeks. They stayed in hotels that quickly became central spaces for establishing scientific networks in an ephemeral but distinct intellectual landscape. Those networks would continue through the season, and often for years after, as Egyptologists built their careers – some more successful or more long-lasting than others. The Hotel du Nil was the hotel that launched a number of careers, from Wallis Budge and Flinders Petrie, to Maggie Benson and James Breasted. The community at the du Nil was not as coherent or cohesive as at other Cairo hotels, but the advice and guidance hotel residents received for purchasing antiquities and finding guides was invaluable to their later successes. Harold Jones found his patrons Emma Andrews and Theodore Davis at Shepheard's. He counted on the terraces at both the Continental and Shepheard's to provide him with dining companions – and jobs. Breasted also used the Continental and Shepheard's as central truth spots for gathering information, intelligence, and conversations over dinners and dances – as much as he did not like those places, he and his work benefited from the proximity to other professionals.

On the Nile, the situation for archaeologists shifted again, especially if it was their first journey. Amelia Edwards, Andrews, Davis, and Charles Wilbour used their time on the Nile to meet new and old friends, ready themselves for the sites they were headed to, work, and relax. The Nile boats – both private dahabeahs and larger steamers – were private and temporary spaces, but were not exactly like hotels. Breasted understood this as he made his first journey up the Nile in 1895, and as he continued his work in 1905. He knew that the boat was useful as a truth spot both for training new crew and being a work space once they arrived on site. He also used the boats he sailed on as sites of disciplinary creation as he worked to professionalise himself early on, and the discipline of archaeology much later. As semi-private spaces where visitors had to be invited on board, they were more private than hotels, but more institutional than them as well.

Luxor was a much more privileged space than hotels in Cairo or boats on the Nile. The resources of time and money that people needed in order to arrive there kept many people away. The colonial spaces and spaces for networking and idea sharing were therefore even more privileged than the others. Chapter 4 illustrates the main points of this book in a much more concrete manner than the ones before it. The Luxor Hotel was a crucial meeting spot for archaeologists who arrived there. It was also an institution in which they could and did store their finds until they could ship them down the Nile to the museum in Cairo. It was a truth spot for the Bensons, Carter, Andrews, and Davis to meet, talk, and relax. The Winter Palace was a centre of international relations and international intellectual networks that flourished in this fortress of Western power. The ways in which Carter, Carnarvon, and Breasted recognised the power of the Winter Palace in their work was clear in their use of the hotel and their rooms as their own ephemeral institutions. They held council and meted out edicts and punishments from the Winter Palace, confident in their authority. The dahabeahs moored in Luxor became more like hotels or semi-public dig houses, as officials from the Department of Antiquities, friends, acquaintances, and others arrived requesting an audience with the owner of the boat.

This book has been a cruise, up and down the Nile, to investigate scientific institutions and activities recognised in the history of science as legitimate spaces where science is done. None of these spaces will come as a surprise to any archaeologists or Egyptologists who have worked in Egypt. What is new is the way that this analysis has combined them and argued that the science done there had as much of an impact on the discipline as the work done in museums, university classrooms, exhibitions, and at scholarly conferences.

The spaces and the ways in which they influence thought will not be surprising to those who read and watch fiction set in Egypt. Readers of novels such as Elizabeth Peters' Amelia Peabody series or Agatha Christie's *Death on the Nile* are familiar with these places as important centres of activity.[2] Hotels and steamers were obviously important social places where scholarship, network building, or murder, could and would occur. Peters' Peabody-Emersons frequently solved murders on their field site, but enjoyed the luxuries of Shepheard's Hotel and sailing on their dahabeah. Christie's Poirot was on board the steamer *Karnak* and had to solve one murder, an attempted murder, and theft before the journey was over. It is true that Christie was married to archaeologist Max Mallowan. She also stayed in the Mena House Hotel, on the Giza Plateau, in her youth and it was rumoured that she wrote part of *Death on the Nile* in the Winter Palace Hotel. In the 1999 film *The Mummy*, the main characters board the steamer *Sudan*, a real, well-known Cook's steamer, to get to the fictional city of Hamunaptra.[3] There, they meet their adversaries and discuss their plans before the boat burns down in the middle of the river, leaving one group with all the horses, and the other group on the right side of the river, as they continue their race to Hamunaptra.

I have had to limit the number of characters or else the stories would become too unwieldy. That is to say, I think a lot more work could be done on different periods of Egyptology, different networks of people, and different nationalities of scientists. I ended in 1925 because that year was the end of the height of colonial power for the British in this part of the world. It is true that they were not fully removed from power until 1952, with the January Revolution and the burning of Shepheard's Hotel, but 1925 was the beginning of the end. Further, by 1925, Egyptology as a Western discipline was fully professionalised and would no longer use hotels or Nile travel in quite the same way. The paths to working in the field and to becoming a professional Egyptologist by that time had changed dramatically. By 1925, a university degree was no longer optional: it was a requirement. A professional or personal connection to a well-known practitioner was still very useful, but those were now made in the classroom and not the hotel dining room. The Harold Joneses of the world would not have fared well without having met the Garstangs of the world in the department at Liverpool first. In other words, by 1925, connections made within the hallowed walls of the university or museum were paramount over those made mainly on the terrace.

The problematic legacy that was established by these early networks in exclusive spaces, however, remains. Today, Egyptologists and other Westerners who work in Egypt have fond memories of their times in the old Windsor, in the original rooms and old bar of the Winter Palace, and the renovated spaces in Mena House Hotel. They have met and will continue to meet friends, family, and colleagues in these places for drinks, meals, and conversation. But these spaces do not hold the same power as they once had in the discipline, even though the issues of voices, access, power, and participation unfortunately do.

Notes

1 See Empereur, *Alexandria rediscovered*; Sydney Hervé Aufrère and Jean-Claude Golvin, *L'Égypte Restituée: Tome 3, Sites, temples et pyramides de Moyenne et Basse Égypt* (Arles: Editions Errance, 1991); Chris Naunton, *Searching for the lost tombs of Egypt* (London: Thames and Hudson, 2018).

2 See Peters, *A crocodile on the sandbank*; Agatha Christie, *Death on the Nile* (London: Collins Crime Club, 1937).

3 Stephen Sommers, Lloyd Fonvielle, Kevin Jarre, Brendan Fraser, Rachel Weisz, John Hannah, Arnold Vosloo, Jonathan Hyde, and Kevin J. O'Connor. *The mummy* (Universal City, CA: Universal, 1999).

Appendix: cast of characters

Unless otherwise noted, all biographical information has been sourced from Morris Bierbrier, *Who was Who in Egyptology* (London: Egypt Exploration Society, 2012). Entries are encylopaedic and in alphabetical order, so no page numbers are given here.

Adriani, Achille (1905–82) Adriani was an Italian archaeologist who was Keeper of the Greco-Egyptian Museum in Alexandria after Breccia, from 1932–40. Then again from 1948–52 after service in the Second World War and as Inspector of Antiquities in Rome.

Allenby, Edmund (1861–1936) General Sir Allenby was the first Viscount Allenby of Megiddo, made so because of his success in battle at the site of Megiddo in Palestine during the First World War. After the War, he was moved to Cairo where he became special high commissioner there. Breasted wrote that Allenby was, essentially, 'king of these lands', and that was how he administered the country. Allenby used Breasted's journey through Mesopotamia in 1919–20 as an intelligence gathering mission, to see what the peoples of Mesopotamia wanted from British rule. The answer was simple: they wanted independence. Allenby did not listen.[1]

Andrews, Emma B. (1837–1922) Andrews was an American archaeologist who travelled up and down the Nile with Theodore Davis from 1889 to 1914. Their first trip to Egypt was in 1887. It is because of Andrews' diaries, now in the Metropolitan Museum of Art's Egyptian Art Department's archives, that we have most of the information about their time in Egypt and their excavations in the Valley of the Kings. These diaries are being transcribed and digitised by Sarah Ketchley. Andrews gave objects to the Museum of Fine Arts, Boston and to the MMA. She left $25,000 to the MMA upon her death.

Benson, Edward F. (1867–1940) Benson was a British novelist and archaeologist. He worked in Athens for the British School of Archaeology from 1892–95 and in Egypt with his sister, Margaret, from 1895–97. He also worked with David Hogarth in that time. After his time as an archaeologist, he continued to travel but lived, largely, in England, where he published over ninety books.[2]

Benson, Margaret (1865–1916) Benson was the first woman to have official permission to excavate in Egypt. Working with her brother, Fred, and her partner, Janet (Nettie) Gourlay, they excavated at the Temple of Mut in Karnak for three seasons, 1895–97. Their publication *The Temple of Mut in Asher* was a best-seller. She was chronically ill, which had an impact on the work she could do in Egypt. She spent much of the rest of her life in a relationship with Janet Gourlay, until Gourlay died in 1912. Upon Benson's death in 1916, she left some of her collections to her nephew, Stewart McDowall. These ended up in the Winchester College Museum. Some of her collections were sold by Christie's in 1972.

Botti, Giuseppe (1853–1903) Botti was an Italian archaeologist who became the first director of the Greco-Egyptian Museum in Alexandria in 1891. He excavated a number of sites in Alexandria, the artefacts from which now populate the museum.

Breasted, James (1865–1935) Breasted was an American Egyptologist who was the first American to earn a PhD in the subject, under Adolf Erman in Berlin in 1894. He founded the Oriental Institute at the University of Chicago and his trips to Egypt from 1894–1935 helped to populate the Institute, along with a number of other institutions in the US, with artefacts from the Nile Valley. He died in New York in 1935. Upon his death, the attending physicians signed an affidavit stating that Breasted did not die from the curse of the mummy of King Tutankhamun, whose tomb he helped to clear, but of a strep infection.

Breccia, Evaristo (1876–1967) Breccia was an Italian archaeologist who became the Director of the Greco-Egyptian Museum in Alexandria after the death of Botti in 1904. He excavated in Alexandria for almost three decades, finding a number of the most intact sites that still draw visitors today. He left the museum in late 1931. After a further career as a professor of classical antiquity in Italy, Breccia died in Rome in 1967.

Brunton, Guy (1878–1948) Brunton was a British Egyptologist who became interested in Egypt as a young child after he read Edwards' *A Thousand Miles up the Nile*. He trained at UCL with Flinders Petrie and Margaret Murray. He excavated with Petrie before the First World War. After the War he returned to Egypt. He worked closely with his wife, Winifred Brunton, on a number of publications. He died in the Transvaal, and his library remains in Johannesburg, South Africa.

Brunton, Winifred (1880–1959) Brunton (née Newberry) was a British artist who trained with Flinders Petrie and Margaret Murray at UCL. She painted detailed watercolours of Egyptian works and tombs, and contributed a number of paintings to her husband's excavation reports. She also published two popular books with a series of her illustrations in them, *Kings and Queens of Ancient Egypt* (1924) and *Great Ones of Ancient Egypt* (1929). Some of her paintings are in the Griffith Institute at Oxford University.

Budge, E. A. Wallis (1857–1934) Budge was a British Egyptologist who was Keeper of the British Museum from 1894–1924. He was Assistant Keeper from 1883–92 and Acting Keeper from 1892–94. He trained at Cambridge and travelled to Egypt for decades. He was knighted in 1920. He published over 140 books and articles, making him one of the most prolific, but not most accurate, Egyptologists. Upon his death he left his library to Christ's College, Cambridge.

Carnarvon, Earl of (see **Herbert**)

Carter, Howard (1874–1939) Carter was a British artist and Egyptologist. He is best known for uncovering the entrance to the tomb of Tutankhamun in 1922. Carter trained with Percy Newberry, beginning when he was seventeen years old. Later he worked with Flinders Petrie at Amarna in 1892 and Edouard Naville from 1893–99. He supervised the excavations of Theodore Davis and Emma Andrews in the Valley of the Kings from 1902–04, when he became the Inspector of Lower Egypt. In 1905 he got into an argument with French tourists at Saqqara, known as the Saqqara Affair, which got him transferred to the Delta region. He began working with Carnarvon in 1907 and it was under his patronage that Carter uncovered Tutankhamun's tomb in 1922. These records are in the Griffith Institute at Oxford University. He died in London.

Davies, Norman de Garis (1865–1941) Davies was a British Egyptologist who depended so much on the work of his wife, Nina (née Cummings, 1881–1965), that it is almost impossible to separate their contributions to Egyptology. After a brief career as a Unitarian minister, he worked with Petrie in 1898 and then other sites for the EEF. He and Nina got married in 1907, and her work as an accomplished and talented artist became entwined with Norman's work. His notebooks and a number of letters, along with her letters and paintings, are at the Griffith Institute.

Davis, Theodore (1838–1915) Davis was an American businessman and patron to Egyptology around the turn of the twentieth century. Wealthy from corrupt legal practices, including with the Boss Tweed Tammany Hall scandals in New York City in the 1860s, Davis needed to spend his money on something philanthropic. With Emma Andrews, he sponsored Percy Newberry and others to excavate in and around Luxor and Thebes in Upper Egypt from around 1900 to 1913. He held the concession for the Valley of the Kings from 1903 to 1914. He and Andrews were responsible for the discovery or clearing of more than twenty tombs in the Valley. Like Andrews, he gave objects to the Museum of Fine Arts, Boston and the

MMA, where most of his private collection ended up after his death and contestation of his will.

Edwards, Amelia (1831–92) Edwards was a novelist, traveller, patron, and founder of the Egypt Exploration Fund. After she visited Egypt in 1873–74, the rest of her life was devoted to the excavation and preservation of Egypt's history. This was problematic, as her Fund depended on British control of Egypt as a colony. She trained Kate Bradbury, Emily Paterson, and other women in administrating the Fund. Her bequest to UCL upon her death founded the first university department of Egyptology in Britain. The Edwards Chair of Egyptology was first occupied by Flinders Petrie, and remains a prestigious position at UCL. Her library and a lot of her personal collection also went to UCL. She died in the home she shared with her partner Ellen Braysher in 1892 and the two are buried together in Henbury Churchyard. Their shared monument there is Grade 2 Listed by Historic England as an important LGBTQ+ landmark.

Forster, Edward M. (1879–1970) Forster was a British novelist and essayist who moved to Alexandria for a brief period during the First World War. He worked for the Red Cross, searching for missing soldiers. While there, he wrote a guidebook about Alexandria, and a series of fictionalised stories about his own life in Alexandria. He wrote numerous other novels and essays later in life.[3]

Garstang, John (1876–1956) Garstang was a British archaeologist who founded the Liverpool Institute of Archaeology in 1904. He trained at Oxford in mathematics, but began excavating Roman sites in England after he left Oxford in 1899 and went to Egypt in 1900. He worked a number of sites while holding the position of Professor of Methods and Practice in Archaeology at Liverpool. His records are still at Liverpool, and the Garstang Museum of Archaeology at Liverpool is named after him.

Golénischeff, Vladimir S. (1856–1947) Golénischeff was a Russian Egyptologist who sold his collection of Egyptian artefacts to the Moscow Museum in 1909. After the Russian Revolution in 1917 he did not return to Russia. He became Professor of Egyptology at Cairo from 1924–29. He lived in Nice for the rest of his life and died there in 1947.

Gourlay, Janet (Janetta) (1863–1912) Gourlay was a British excavator who worked with Margaret Benson on the Temple of Mut in Karnak from 1896–97, and then excavated with Percy Newberry in 1900–01. She trained with Flinders Petrie at UCL, so offered Benson some expertise that she did not already have. Not much more is known about her, except that she had a loving relationship with Margaret Benson for the rest of her life. She died unmarried, but not alone, in 1912.

Herbert, George Edward Stanhope Molyneaux (1866–1923) Known better as the 5th Earl of Carnarvon, he was a British patron and collector. He began wintering in Egypt in 1903, a couple of years after a motor accident weakened his health. From 1907 he hired Howard Carter to excavate for him in Thebes. In 1922, the final season of his support, Carter uncovered the tomb of Tutankhamun. In early 1923, Carnarvon died in his hotel room at the Continental in Cairo from an infected mosquito bite on his cheek causing his blood to be septic. Many thought that this was the curse of the mummy, but it was simply a bacterial infection that could not be treated effectively.

Hogarth, David G. (1862–1927) Hogarth was an English classicist and excavator. He was educated at Oxford and became a field archaeologist early on. From 1894 to 1896 he excavated for the EEF at a few sites, including Alexandria. During the First World War he was the Director of the Arab Bureau in Cairo and worked with Gertrude Bell and T. E. Lawrence gathering and disseminating intelligence about peoples in the Middle East for British Intelligence.

Jones, E. Harold (1877–1911) Jones was a British artist and excavator who first came to Egypt looking for relief from his tuberculosis symptoms. He started working with Garstang in 1904, but soon joined Theodore Davis and Emma Andrews in their work in the Valley of the Kings. He worked with them for four years. He died of tuberculosis in 1911 and is buried in Luxor.

Kamal, Ahmed (1851–1923) Kamal Pasha was an Egyptian Egyptologist who was the first Egyptian to become both an archaeologist and Egyptologist. He became a secretary-interpreter to the Antiquities Service, then became Assistant Curator of the Museum in Cairo. He researched the development of languages but died before his work was published. He was made Pasha shortly before his death in 1923.

Lane, Jenny (b. 1835) Lane was a British ladies' maid to Lucy Renshaw and Amelia Edwards on their trip up the Nile in 1873–74. Her diaries are kept at the Griffith Institute and reveal a side of the journey that was not presented in Edwards' famous travelogue.[4]

Maspero, Gaston (1846–1916) Maspero was a French Egyptologist who was trained by Auguste Mariette from a young age. He had a long and storied career. In the end, he was appointed Director of the Bulaq Museum in 1881–86, making him also the Director of the Antiquities Service. He went back to France from 1886–99 to teach at the Collège de France. He then returned to Egypt in 1899 to be the Director of the Antiquities Service until 1914, when he left for France due to illness. He died in Paris in 1916.

Mohassib, Mohammed (1843–1928) Mohassib was a trusted antiquities dealer in the Luxor area. He started his career as a donkey boy for Lady Lucie Duff Gordon in the 1860s, but would have been over the age of twenty when they met. He practised English with her, as well. He opened his shop in Luxor in the 1880s and quickly became a trusted dealer and sold objects to most of the people in this book, including Andrews, Budge, Breasted, Carter, Davis, Maspero, and Wilbour.

Murray, Margaret (1863–1963) Murray was a British Egyptologist who was the first university-trained woman Egyptologist in Britain. She trained under Flinders Petrie at UCL from 1894; by 1898 she was teaching classes, but never earned her degree. During her forty years at UCL, she developed the first two-year training programme with the aim of preparing students to go into the field. She also managed the museum collections and taught at least five classes per term. She worked on site with Petrie only a few times because he needed her to be in London while he was gone. She loved to pass out chocolates to her students during class while she lectured on folklore and witchcraft.

Newberry, Percy (1868–1949) Newberry was a British Egyptologist who may be best known for his prolific correspondence that is now held in the Griffith Institute. He began work with R. S. Poole and the EEF in 1884 and was associated with them for the rest of his career. He started work with Petrie on site reports, then quickly moved to independent work. He superintended excavations for Margaret Benson, Theodore Davis, Emma Andrews, and more in Thebes. He was on faculty at Liverpool from 1906–19 and Professor of Ancient History and Archaeology at Cairo University from 1929 to 1933.

Petrie, William Mathew Flinders (1853–1942) Petrie was a British Egyptologist who largely taught himself, measuring Stonehenge with his father before moving to the Giza Plateau in 1880. He quickly became a protégé of Amelia Edwards, and because of his work with her and the EEF, he was chosen to be the first Edwards Chair of Egyptology at UCL in 1894. He remained at UCL until 1935, excavating in Egypt almost every season for seventy years. He could not have been as prolific without the erstwhile assistance of an army of students and assistants, both men and women, including Margaret Murray and his wife, Hilda Petrie. Petrie was a fervent eugenicist, sending masses of skulls and skeletons back to Francis Galton and Karl Pearson at UCL for decades. He died in Jerusalem in 1942 and had his head sent to the Royal College of Surgeons for eugenic study.

Petrie, Hilda Mary Isabel (1871–1956) Petrie (née Urlin) was a British Egyptologist trained by Flinders Petrie on their honeymoon in 1897. She did a lot of the record keeping and correspondence on site as well as managing crew members and the domestic side of excavation life. She and Flinders had two children, but Hilda rarely missed an excavation season, often leaving the children in England with nannies. She is the one on whose toil most of Flinders' work depended. She died in London.

Quibell, James E. (1867–1935) Quibell was a British Egyptologist who began working with Flinders Petrie at Koptos (Quft) in 1893. At this site, Quibell helped Petrie train the men in his crew who would become famous around Egypt and the Middle East as the best trained excavators available. Known as Quftis, these crew continue to train each other and are sought out all over for their expertise. He became an inspector in 1899 and served all over Egypt in that capacity until 1914. He was the Keeper of the Egyptian Museum from 1914 to 1923 when he retired. He and his wife, Annie Quibell, an archaeologist herself, met and fell in love over bouts of ptomaine poisoning due to rotten tinned food on Petrie's excavations.

Reisner, George (1867–1942) Reisner was an American Egyptologist who studied with Erman in Berlin; he must have barely overlapped or just missed Breasted. He became Director of the Hearst Egyptian Expedition in 1899 and spent a lot of time excavating on the Giza Plateau, where he found the tomb of Queen Hetepheres. He excavated all over Egypt but much of his work remained unpublished at his death.

Roger, Guthrie (n.d.) Roger, a British excavator, worked only one partial season with the EEF and did not enjoy his time. We only know about him from the few letters he sent to Percy Newberry and to Emily Paterson at the EEF in 1893–94. He was supposed to work with Carter, but, due to problems in getting permission, that did not happen. He hoped for more work in Egypt but ended up having to come home.

Sayce, Archibald (1845–1933) Sayce was a British Assyriologist who spent a lot of time on the Nile with Gaston Maspero and Charles Wilbour. He was educated at Oxford and was Professor of Assyriology there from 1891–1919. He copied a number of inscriptions and was known for his work in scripts. His collection of Egyptian antiquities was bequeathed to the Ashmolean Museum, Oxford, on his death.

Tirard, Helen Mary (1852–1943) Tirard (née Beloe) was a British translator who was an early member of the EEF. She donated thousands of pounds to the efforts of the EEF and served with Emily Paterson on the EEF Committee for decades. She wrote a travelogue of her time on a Cook's steamer and gave numerous public talks to encourage public interest and support in the EEF's work. She translated Erman's *Life in Ancient Egypt*. She became blind in her later years and died in 1943.

Wilbour, Charles (1833–96) Wilbour was an American traveller and patron of Egyptology who was trained by Gaston Maspero in Paris. Wilbour was wealthy due to his connections and activity in the Boss Tweed scandal in the 1860s. He had to leave New York to escape prosecution, but needed to spend his money somewhere. He moved his family to Paris and soon began sailing up and down the Nile, first on the Cairo Museum steamer with Maspero, then on his own dahabeah. He aided Maspero in copying and translating inscriptions, but never published any substantial work. His legacy is most clearly seen in the Wilbour Library and the Egyptian Art collection at the Brooklyn Museum.

Notes

1 Matthew Hughes, 'Allenby, Edmund Henry Hynman, first Viscount Allenby of Megiddo (1861–1936)', *Oxford Dictionary of National Biography* (23 September 2004).
2 Sayoni Basu, 'Benson, Edward Frederic (1867–1940)', *Oxford Dictionary of National Biography* (26 May 2005).
3 Nicola Beauman, 'Forster, Edward Morgan (1879–1970)', *Oxford Dictionary of National Biography* (28 September 2006).
4 Collection J. Lane MSS – Jenny Lane Collection, Griffith Institute, https://archive.griffith.ox.ac.uk/index.php/lane-jenny (accessed 26 July 2021).

Bibliography

Archives

Individual archive documents are referenced in the text.
Egypt Exploration Society (EES), London, England.
Garstang Museum Archives, Liverpool, England.
Griffith Institute of Egyptology, Oxford University, Oxford, UK.
James H. Breasted Correspondence, Archives of the Oriental Institute, University of Chicago, Chicago, USA.
The Metropolitan Museum of Art, Department of Egyptian Art Archives, New York City, New York.
Thomas Cook Archives, Thomas Cook Offices, Peterborough, UK (the archives have since moved to the National Archives at Kew Gardens and may have different references).
Wilbour Papers, Brooklyn Museum, Brooklyn, New York.

Internet sources

Adams, John M. 'A bad dream of New York: The rise, fall, and redemption of Charles E. Wilbour.' n.d. /www.academia.edu/6990369/A_BAD_DREAM_OF_NEW_YORK_The_Rise_Fall_and_Redemption_of_Charles_E_Wilbour.
Brooklyn Museum, *Temple of Mut* www.brooklynmuseum.org/features/mut.
Collection J. Lane MSS – Jenny Lane Collection, Griffith Institute, https://archive.griffith.ox.ac.uk/index.php/lane-jenny.
CPI Inflation Calculator, www.in2013dollars.com/.
Emma B. Andrews Diary Project, www.emmabandrews.org/project/.
'The Epigraphic Survey: The 'Chicago House Method', https://oi.uchicago.edu/research/projects/epi/chicago-house-method.
Humphreys, Andrew. *Grand Hotels of Egypt.* http://grandhotelsegypt.com/.
Islington Studios 'The Camels are Coming.' 1934. www.youtube.com/watch?v=YDuABIATHrc.
Lawrence, Deirdre. 'Wilbour: One man's obsession with Egypt.' *Brooklyn Museum Blog*: www.brooklynmuseum.org/community/blogosphere/2010/03/22/wilbour-one-mans-obsession-with-egypt/.
Measuring Worth, www.measuringworth.com/calculators/exchange/.

National Archives Currency Converter, www.nationalarchives.gov.uk/currency-converter/.

Oxford Dictionary of National Biography, www.oxforddnb.com/.

Peck, William H. *E. F. Benson in Egypt*. http://williamhpeck.org/e_f_benson_in_egypt.

Scalf, Foy D. *The research archives of the Oriental Institute: Introduction and guide*. Chicago: Oriental Institute, 2013. https://oi.uchicago.edu/sites/oi.uchicago.edu/files/uploads/shared/docs/research_archives_introduction&guide.pdf.

Sheppard, Kathleen. '"Constant Companions" and "Intimate Friends": The lives and careers of Maggie Benson and Nettie Gourlay.' *Lady Science* 57 (6 June 2019). www.ladyscience.com/constant-companions-and-intimate-friends/no57.

Teicholz, Tom. 'The British Museum: The problematic yet enduring appeal of Antiquities.' *Forbes* (14 October 2018). www.forbes.com/sites/tomteicholz/2018/10/14/the-british-museum-the-problematic-yet-enduring-appeal-of-antiquitie/#314d6bf74244.

Books and articles

'America's new Diplomatic Agent and Consul General.' *The Egyptian Gazette* (2 September 1893): 2.

'Breasted sails with John D. Jr. on Egypt Tour.' *Chicago Tribune*, 3 January 1929.

Cook's tourists' handbook for Egypt, the Nile, and the Desert. London: Thomas Cook & Son, 1876.

Cook's tourists' handbook for Egypt, the Nile, and the Desert. London: Thomas Cook & Son, 1897.

'Condemns Roosevelt Speech.' *New York Times* (1 April 1910): 4.

'Dahabeah Arrangements.' *Cook's Excursionist and Tourist Advertiser* (16 September 1895): 7.

Proceedings of the Society of Antiquaries in London. London, 1905.

'Upper Egypt Hotels Co. Inauguration of Winter Palace.' *The Egyptian Gazette* (19 January 1907).

Abd el-Gawad, Heba and Alice Stevenson. 'Egypt's dispersed heritage: Multi-directional storytelling through comic art.' *Journal of Social Archaeology* 21:1 (2021): 121–45. DOI: https://doi.org/10.1177%2F1469605321992929.

Abt, Jeffrey. *American Egyptologist: The life of James Henry Breasted and the creation of his Oriental Institute*. Chicago: University of Chicago Press, 2011.

Adams, John. *The millionaire and the mummies: Theodore Davis's Gilded Age in the Valley of the Kings*. New York: St Martin's Press, 2013.

Aldridge, James. *Cairo*. Boston: Little, Brown and Company, 1969.

Allen, Susan. 'Tycoons on the Nile: How American millionaires brought Egypt to America.' In *Lost and now found: Explorers, diplomats and artists in Egypt and the Near East*, eds Neil Cooke and Vanessa Daubney, 71–81. Oxford: ASTENE, 2017.

Aufrère, Sydney Hervé and Jean-Claude Golvin. *L'Égypte Restituée: Tome 3, Sites, temples et pyramides de Moyenne et Basse Égypt*. Arles: Editions Errance, 1991.

Barr, James. *Setting the desert on fire: T. E. Lawrence and Britain's secret war in Arabia, 1916–1918*. New York: W. W. Norton & Co., 2008.

Barr, James. *A line in the sand: Britain, France and the struggle for the mastery of the Middle East*. London: Simon & Schuster, 2011.

Barr, James. *Lords of the desert: The battle between the United States and Great Britain for supremacy in the modern Middle East*. New York: Basic Books, 2018.

Basu, Sayoni. 'Benson, Edward Frederic (1867–1940).' *Oxford Dictionary of National Biography*, 26 May 2005.

Beauman, Nicola. 'Forster, Edward Morgan (1879–1970).' *Oxford Dictionary of National Biography*, 28 September 2006.

Bell, Martha R. 'An armchair excavation of KV 55.' *Journal of the American Research Center in Egypt* 27 (1990): 97–137.

Benson, Arthur C., ed. *Life and letters of Maggie Benson*. London: John Murray, 1917.

Benson, Margaret and Janet Gourlay. *The Temple of Mut in Asher*. London: John Murray, 1899.

Berger, Molly. *Hotel dreams: Luxury, technology, and urban ambition in America, 1829–1929*. Baltimore: Johns Hopkins University Press, 2011.

Berghoff, Hartmut, Barbara Korte, Ralf Schneider, and Christopher Harvie, eds. *The making of modern tourism: The cultural history of the British experience, 1600–2000*. New York: Palgrave MacMillan, 2002.

Bierbrier, Morris. *Who was who in Egyptology*. London: Egypt Exploration Society, 2012.

Bird, Michael. *Samuel Shepheard of Cairo: A portrait*. London: Michael Joseph, 1957.

Blouin, Katherin. *Triangular landscapes: Environment, society, and the state in the Nile Delta under Roman rule*. Oxford: Oxford University Press, 2014.

Botti, Giuseppe, ed. *Bulletin de la Société Archéologique d'Alexandrie*, 1. Alexandria: La Société, 1898.

Botti, Giuseppe, ed. *Plan de la ville d'Alexandrie à l'époque ptolémaique Monuments et localités de l'ancienne Alexandrie; Mémoire présenté à la société archéologique*. Alexandria: La Société, 1898.

Breasted, Charles. *Pioneer to the past: The story of James Henry Breasted archaeologist*. Chicago: The Oriental Institute, 1943.

Breasted, James. *A History of Egypt from the earliest times to the Persian Conquest*, 2nd edn. London: Hodder and Stoughton, 1927.

Breccia, Evaristo. *Alexandrea ad Aegyptum: A guide to the ancient and modern town and to its Graeco-Roman Museum*. Bergamo: Instituto Italiano d'Arti Grafiche, 1922.

Brendon, Piers. *Thomas Cook: 150 years of popular tourism*. London: Secker & Warburg, 1991.

Brier, Bob. *Cleopatra's Needles: The lost obelisks of Egypt*. London: Bloomsbury Egyptology, 2016.

Brocklehurst, Marianne. *Miss Brocklehurst on the Nile: Diary of a Victorian traveller in Egypt*. Cheshire: Millrace, 2004.

Brodrick, Mary. *A handbook for travellers in Lower and Upper Egypt*. London: John Murray, 1900.

Browne, Janet. 'Corresponding naturalists.' In *The age of scientific naturalism: Tyndall and his contemporaries*, eds Bernard Lightman and Michael S. Reidy, 157–69. London: Pickering and Chatto, 2014.

Brunton, Winifred. *Kings and Queens of ancient Egypt. Portraits by Winifred Brunton. History by eminent Egyptologists, etc.* London: Hodder and Stoughton, 1924.

Brunton, Winifred. *Great ones of ancient Egypt.* London: Hodder and Stoughton, 1929.

Budge, E. A. Wallis. *Cook's handbook for Egypt and the Sudan.* London: Thomas Cook & Son, 1905.

Budge, E. A. Wallis. *The Nile: Notes for travellers in Egypt*, 9th edn. London: Thomas Cook & Son, 1905.

Budge, E. A. Wallis. *By Nile and Tigris: A narrative of journeys in Egypt and Mesopotamia on behalf of the British Museum between the years 1886 and 1913.* London: John Murray, 1920.

Capart, Jean, ed. *Travels in Egypt (December 1880 to May 1891): Letters of Charles Edwin Wilbour.* Brooklyn: Brooklyn Institute of Arts and Sciences, 1936.

Carruthers, William, ed. *Histories of Egyptology: Interdisciplinary measures.* London: Routledge, 2014.

Carruthers, William. 'Credibility, civility, and the archaeological dig house in mid-1950's Egypt.' *Journal of Social Archaeology* 19:2 (2019): 255–76.

Carruthers, William and Stéphane Van Damme, eds. *History of Science, Special Issue: Disassembling archaeology, reassembling the modern world* 55:3 (2017).

Carter, H. and A. C. Mace. *The tomb of Tut-ankh-Amen, discovered by the late Earl of Carnarvon and Howard Carter, Vol. 1.* New York: George H. Doran, Co., 1923.

Carter, H. and A. C. Mace. *The tomb of Tut-ankh-Amen, discovered by the late Earl of Carnarvon and Howard Carter, Vol. 2.* New York: George H. Doran, Co., 1927.

Challis, Debbie. *The archaeology of race: The eugenic ideas of Francis Galton and Flinders Petrie.* London: Bloomsbury, 2013.

Christie, Agatha. *Death on the Nile.* London: Collins Crime Club, 1937.

Colla, Elliott. *Conflicted antiquities: Egyptology, Egyptomania, Egyptian modernity.* Durham, NC: Duke University Press, 2007.

Creighton, Rev. O. *With the twenty-ninth division in Gallipoli: A chaplain's experiences.* London: Longman's, Green & Co., 1916.

Daressy, George. 'Catalogue of the jewels and precious objects of Setuî II and Tauosrît found in the unnamed tomb.' In *The tomb of Siphtah*, ed. Theodore M. Davis. 33–46. London: Duckworth, 1908.

Davis, Theodore M. *The tomb of Siphtah; The Monkey tomb and the Gold tomb.* London: Archibald Constable & Co., 1908.

Davis, Theodore M. *The tomb of Queen Tiyi.* London: Duckworth, 1908.

Davis, Theodore M. 'The finding of the tomb of Queen Tîyi.' In *The tomb of Queen Tîyi*, 1–5. London: Duckworth, 1908.

Davis, Theodore M. *The tombs of Harmhabi and Touatânkhamanou: The discovery of the tombs.* London: Constable and Co., 1912.

Delany, C. *A son to Luxor's sand: A commemorative exhibition of Egyptian art from the collections of the British Museum and Carmarthen Museum.* Dyfed City Council, 1986.

Denon, Vivant. *Voyage dans la Basse et la Haute Egypte pendant les campagnes du General Bonaparte*. Paris: P. Didot, 1802.

Díaz-Andreu, Margarita. *A history of archaeological tourism: Pursuing leisure and knowledge from the eighteenth century to World War II*. Cham, Switzerland: Springer, 2019.

Dobson, Eleanor. 'A tomb with a view: Supernatural experiences in the late nineteenth century's Egyptian hotels.' In *Anglo-American travelers and the hotel experience in nineteenth-century literature: Nation, hospitality, and travel writing*, eds Monika M. Elbert and Susanne Schmid, 89–105. London: Routledge, 2017.

Doyon, Wendy. 'The history of archaeology through the eyes of Egyptians.' In *Unmasking ideology in imperial and colonial archaeology: Vocabulary, symbols, and legacy*, eds Bonnie Effros and Guolong Lai, 173–200. Los Angeles: Cotsen Institute of Archaeology Press, 2018.

Drower, Margaret, ed. *Letters from the desert: The correspondence of Flinders and Hilda Petrie*. Oxford: Aris & Phillips, 2004.

Duncan, James. 'Dis-Orientation: On the shock of the familiar in a far-away place.' In *Writes of passage: Reading travel writing*, eds James Duncan and Derek Gregory, 151–63. London: Taylor & Francis, 1998.

Edwards, Amelia. *A thousand miles up the Nile*. London: Longmans, Green, 1877.

El-Din, Morsi Saad. 'Introduction.' In *Alexandria: The site & the history*, ed. Gareth L. Steen, 9–18. New York: NYU Press, 1993.

Elliot Smith, Grafton. 'A note on the estimate of the age attained by the person whose skeleton was found in the tomb.' In *The tomb of Queen Tîyi*, ed. Theodore M. Davis, xxiii–xxiv. London: Duckworth, 1908.

Elshahed, Mohamed. *Cairo since 1900: An architectural guide*. Cairo: AUC Press, 2019.

Empereur, Jean-Yves. *Alexandria rediscovered*. London: British Museum Press, 1998.

Erman, Adolf. *Aegypten und aegyptisches Leben*. Tübingen: H. Laupp'sche Buchhandlung, 1885.

Erman, Adolf, transl. H. M. Tirard, *Life in Ancient Egypt*. New York: Macmillan & Co., 1894.

Erman, Adolf and Hermann Grapow, eds. *Wörterbuch der Aegyptischen Sprache*. Berlin: Akademie-Verlag, 1926–61.

Farrell, Michael P. *Collaborative circles: Friendship dynamics and creative work*. Chicago: University of Chicago Press, 2001.

Finnegan, Ruth, ed. *Participating in the knowledge society: Researchers beyond the university walls*. London: Palgrave MacMillan, 2005.

Forster, E. M. *Pharos and Pharillon*. New York: Alfred A. Knopf, 1923.

Forster, E. M. *Alexandria: A history and a guide*. Alexandria: Whitehead Morris Limited, 1938.

Freedman, Paul. *Ten restaurants that changed America*. New York: Liveright, 2016.

Fyfe, Aileen and Bernard Lightman, eds. *Science in the marketplace: Nineteenth-century sites and experiences*. Chicago: University of Chicago Press, 2007.

Gardiner, Alan. *My working years*. Privately published, 1963.

Garnett, Anna. 'John Rankin and John Garstang: Funding Egyptology in a pioneering age.' In *Forming material Egypt: Proceedings of the International Conference,*

London, 20–21 May, 2013, eds P. Piacentini, C. Orsenigo, and S. Quirke, 95–104. Milan: Pontremoli, 2013–14.

Gieryn, Thomas. *Truth spots: How places make people believe*. Chicago: University of Chicago Press, 2018.

Glänzel, W. and A. Schubert. 'Analysing scientific networks through co-authorship.' In *Handbook of quantitative science and technology research: The use of publication and patent statistics in studies of S&T systems*, eds H. F. Moed, W. Glänzel, and U. Schmoch, 257–66. New York: Kluwer Academic Publishers, 2005.

Goldhill, Simon. *A very queer family indeed: Sex, religion, and the Bensons in Victorian Britain*. Chicago: University of Chicago Press, 2016.

Goode, James. *Negotiating for the past: Archaeology, nationalism, and diplomacy in the Middle East, 1919–1941*. Austin: University of Texas Press, 2007.

Goodman, Dena. *The republic of letters: A cultural history of the French Enlightenment*. Ithaca: Cornell University Press, 1994.

Gregory, Derek. 'Scripting Egypt: Orientalism and the cultures of travel.' In *Writes of passage: Reading travel writing*, eds James S. Duncan and Derek Gregory, 114–50. New York: Routledge, 1999.

Gregory, Derek. 'Emperors of the gaze: Photographic practices and productions of space in Egypt, 1839–1914.' In *Picturing place: Photography and the geographical imagination*, eds Joan M. Schwartz and James R. Ryan, 195–225. London: I. B. Tauris, 2003.

Grewal, Inderpal. 'The guidebook and the museum.' In *Home and harem: Nation, empire and the cultures of travel*, 85–130. Durham, NC: Duke University Press, 1996.

Haag, Michael. *Alexandria: City of memory*. New Haven: Yale University Press, 2004.

Haag, Michael, ed. *An Alexandria anthology: Travel writing through the centuries*. Cairo: AUC Press, 2014.

Hall, H. R. ed. *Murray's, A handbook for Egypt and the Sudan*. London: John Murray, 1907.

Harer, Jr., W. Benson. 'The Drexel Collection: From Egypt to the Diaspora.' In *Servant of Mut: Studies in honor of Richard A. Fazzini*, ed. Sue H. D'Auria, 111–19. Leiden: Brill, 2008.

Hilgartner, Stephen. *Science on stage: Expert advice as public drama*. Stanford: Stanford University Press, 2000.

Hinrichsen, Alex. *Baedeker's travel guides, 1832–1990*, 2nd edn, transl. Åke Nilson. Bevern: Verlag Ursula Hinrichsen, 1991.

Hogarth, David George and Edward Fredrick Benson. 'Report on prospects of research in Alexandria: with note on excavations in Alexandrian cemeteries.' *Archaeological Report (Egypt Exploration Fund) 1894–1895*, 1–33. London: Macmillan, 1895.

Holmes, Mary J. 'Street Life in Egypt.' *St. Louis Post Dispatch* (9 February 1890): 24.

Hughes, Matthew. 'Allenby, Edmund Henry Hynman, first Viscount Allenby of Megiddo (1861–1936).' *Oxford Dictionary of National Biography*, 23 September 2004.

Hulme-Beaman, Ardern G. *Travels without Baedeker*. New York: John Lane, 1913.

Humphreys, Andrew. *Grand hotels of Egypt in the golden age of travel*. Cairo: AUC Press, 2010.

Humphreys, Andrew. *On the Nile in the golden age of travel*. Cairo: AUC Press, 2015.

Ibrahim, Tarek. *Shepheard's of Cairo: The birth of the Oriental grand hotel*. Wiesbaden: Reichert Verlag: 2019.

James, Kevin J., A. K. Sandoval-Strausz, Daniel Maudlin, Maurizio Peleggi, Cédric Humair, and Molly W. Berger. 'The hotel in history: evolving perspectives.' *Journal of Tourism History* 9:1 (2017): 92–111.

James, T. G. H. *Howard Carter: The path to Tutankhamun*. London: I. B. Tauris, 2001.

Janssen, Rosalind. *The first hundred years: Egyptology at University College London, 1892–1992*. London: UCL Press, 1992.

Kark, Ruth. *American consuls in the Holy Land, 1832–1914*. Jerusalem: The Hebrew University, 1994.

Ketchley, Sarah. '"Witnessing the 'Golden Age": The Diaries of Mrs. Emma B. Andrews.' *KMT* (December 2020): 33–43.

Larson, John A. ed. *Letters from James Henry Breasted to his family, August 1919–July 1920: Letters home during the Oriental Institute's first expedition to the Middle East*. Oriental Institute Digital Archives, No. 1. Chicago: Oriental Institute, 2010.

Larson, John A. 'Introduction.' In *Letters from James Henry Breasted to his family, August 1919–July 1920 Letters home during the Oriental Institute's First Expedition to the Middle East*. Oriental Institute Digital Archives, No. 1. ed. John A. Larson, 11–27. Chicago: Oriental Institute, 2010.

Lehnert, Isolde. '"Let's have a beer at Gorff's!"' In *Journeys erased by time: The rediscovered footprints of travellers in Egypt and the Near East*, ed. Neil Cooke, 115–32. Oxford: Archaeopress, 2019.

Livingstone, David. *Putting science in its place: Geographies of scientific knowledge*. Chicago: University of Chicago Press, 2003.

Livingstone, David. 'Science, site and speech: Scientific knowledge and the spaces of rhetoric.' *History of the Human Sciences* 20:2 (2007): 71–98.

Livingstone, David and Charles Withers. *Geographies of nineteenth century science*. Chicago: University of Chicago Press, 2011.

Lorimer, Norma. *By the waters of Sicily*. New York: James Pott & Co., 1901.

Lorimer, Norma. *By the waters of Carthage*. New York: James Pott, & Co., 1906.

Lorimer, Norma. *By the waters of Egypt*. London: Methuen & Co, 1909.

Lorimer, Norma. *By the waters of Germany*. London: S. Paul, 1914.

Loti, Pierre. *Egypt (La Mort de Philae)*. New York: Duffield & Co., 1910.

Mackenzie, John. *Orientalism: History, theory and the arts*. Manchester: Manchester University Press, 1995.

Mairs, Rachel and Maya Muratov. *Archaeologists, tourists, interpreters: Exploring Egypt and the Near East in the late 19th–early 20th centuries*. London: Bloomsbury, 2015.

Mak, Lanver. *The British in Egypt: Community, crime and crises 1882–1922*. London: I. B. Tauris, 2012.

Manley, Deborah. *A Cairo anthology: Two hundred years of travel writing*. Cairo: AUC Press, 2013.

Manley, Deborah, ed. *Women travelers on the Nile*. Cairo: AUC Press, 2016.

Marcus, Sharon *Between women: Friendship, desire, and marriage in Victorian England*. Princeton: Princeton University Press, 2007.

Mariette, Auguste. *Karnak: étude topographique et archéologique avec un appendice comprenant les principaux textes hiéroglyphiques découverts ou recueillis pendant les fouilles exécutées à Karnak*. Leipzig: J.C. Hinrichs, 1875.

Marino, Elisabetta. 'Three British women travelers in Egypt: Sophia Lane Poole, Lucie Duff Gordon, and Emmeline Lott.' In *The legacy of the grand tour: New essays on travel, literature, and culture*, ed. Lisa Coletta, 51–70. Lanham: Rowman & Littlefield, 2015.

Mecham, G. W. 'Tut's Tomb becomes Mecca for Tourists.' *The Chicago Daily News* (6 February 1924), 2.

Mecham, G. W. 'Breasted in Protest over Tut Tomb Ban.' *The Chicago Daily News* (14 February 1924), 2.

Mecham, G. W. 'Tut Tomb Discoverer Barred Out By Police.' *The Chicago Daily News* (15 February 1924), 2.

Mickel, Allison. *Why those who shovel are silent: A history of local archaeological knowledge and labor*. Louisville: University Press of Colorado, 2021.

Miller, David Phillip. 'Method and the "micropolitics" of science: The early years of the Geological and Astronomical Societies of London.' In *The politics and rhetoric of scientific method*, eds John A. Schuster and Richard Yeo, 227–57. Boston: D. Reidel, 1986.

Mitchell, Timothy. *Colonizing Egypt*. Cambridge: Cambridge University Press, 1988.

Mitchell, Timothy. *Rule of experts: Egypt, techno-politics, modernity*. Berkeley: University of California Press, 2002.

Moon, Brenda. *More usefully employed: Amelia B. Edwards, writer, traveler and campaigner for ancient Egypt*. London: Egypt Exploration Society, 2006.

Morgan, Colleen and Daniel Eddisford. 'Dig houses, dwelling, and knowledge production in archaeology.' *Journal of Contemporary Archaeology* 2:1 (2015): 169–93.

Morris, Richard E. 'The Victorian "Change of Air" as medical and social construction.' *Journal of Tourism History* 10:1 (2018): 49–65.

Naunton, Chris. *Searching for the lost tombs of Egypt*. London: Thames and Hudson, 2018.

Naunton, Chris. *Egyptologists' notebooks: The golden age of Nile exploration in words, pictures, plans and letters*. London: Thames & Hudson, 2020.

Naylor, Simon. 'The field, the museum and the lecture hall: the spaces of natural history in Victorian Cornwall.' *Transactions of the Institute of British Geographers* 27:4 (2002): 494–513.

Naylor, Simon. 'Introduction: historical geographies of science – places, contexts, cartographies.' *British Journal for the History of Science* 38:1 (2005): 1–12.

Nelson, Nina. *Shepheard's Hotel*. London: Barrie and Rockliff, 1960.

Newberry, Percy E. 'B. The Archaeological Survey of Egypt: Mr. Newberry's Work, 1892–93.' *Egypt Exploration Fund, Archaeological Report 1892–1893*, 9–15. London: Kegan Paul, 1893.

Newberry, Percy E. 'Notes and News.' *Journal of Egyptian Archaeology* 14:1/2 (May 1928): 184.

Newberry, Percy E. 'Howard Carter.' *Journal of Egyptian Archaeology* 25:1 (1939): 67.

Newman, M. E. 'The structure of scientific collaboration networks.' *Proceedings of the National Academy of Sciences of the United States of America* 98:2 (2001): 404–9.

Peel, Victoria and Anders Sørensen. *Exploring the use and impact of travel guidebooks*. Toronto: Channel View Publications, 2016.

Peters, Elizabeth. *Crocodile on the sandbank*. New York: Dodd, Mead, 1975.

Peters, Elizabeth and Kristen Whitbread, eds. *Amelia Peabody's Egypt: A compendium*. New York: William Morrow, 2003.

Petrie, W. M. Flinders. 'A Digger's Life.' *The English Illustrated Magazine* (March 1886): 440–1.

Petrie, W. M. Flinders. *Seventy years in archaeology*. New York: Henry Holt, 1931.

Petrie, W. M. Flinders and J. E. Quibell. *Naqada and Ballas 1895*. London: Bernard Quaritch, 1896.

Pirie, Gordon. 'Incidental tourism: British Imperial air travel in the 1930s.' *Journal of Tourism History* 1:1 (2009): 49–66.

Pincus, Steve. '"Coffee Politicians Does Create": Coffeehouses and Restoration political culture.' *The Journal of Modern History* 67:4 (1995): 807–34.

Playfair, Robert. *Handbook to the Mediterranean*, 2nd edn. London: John Murray, 1882.

Quirke, Stephen. *Hidden hands: Egyptian workforces in Petrie excavation archives, 1880–1924*. London: Bloomsbury, 2010.

Qureshi, Sadiah. *Peoples on parade: exhibitions, empire, and anthropology in nineteenth century Britain*. Chicago: University of Chicago Press, 2011.

Ramadan, Abdel/Azim. 'Alexandria: French expedition to the modern age.' In *Alexandria: the site & the history*, ed. Gareth L. Steen, 109–26. New York: NYU Press, 1993.

Raymond, André. transl. Willard Wood. *Cairo*. Cambridge: Harvard University Press, 2000.

Reeve, Charles McCormick. *How we went and what we saw: A flying trip through Egypt, Syria, and the Aegean Islands*. New York: G. P. Putnam's Sons, 1891.

Reeves, C. Nicholas. *The Valley of the Kings*. Kegan Paul, 1990.

Reeves, Nicholas and John H. Taylor. *Howard Carter: Before Tutankhamun*. New York: Harry Abrams, 1993.

Reid, Donald Malcolm. 'The Urabi revolution and the British conquest, 1879–1882.' In *The Cambridge history of Egypt (Volume 2)*, ed. M. W. Daly, 217–38. Cambridge: Cambridge University Press, 1999.

Reid, Donald Malcolm. *Whose Pharaohs?: Archaeology, museums, and Egyptian national identity from Napoleon to World War I*. Los Angeles: University of California Press, 2002.

Reid, Donald Malcolm. *Contesting antiquity in Egypt: Archaeologies, museums & the struggle for identities from World War I to Nasser*. Cairo: American University in Cairo Press, 2015.

Reynolds-Ball, Eustace Alfred. *Cairo of to-day; a practical guide to Cairo and the Nile*. London: A and C Black, 1916.

Riggs, Christina. *Unwrapping ancient Egypt*. London: Bloomsbury, 2014.

Riggs, Christina. *Photographing Tutankhamun: Archaeology, ancient Egypt, and the archive*. London: Routledge, 2019.

Roberts, Julia, Kathleen Sheppard, Ulf Hansson, Jonathan Trigg, eds. *Communities and knowledge production in archaeology*. Manchester: Manchester University Press, 2020.

Rodenbeck, Max. *Cairo: The city victorious*. New York: Vintage Books, 1998.

Rudwick, Martin J. S. *The Great Devonian Controversy: The shaping of scientific knowledge among gentlemanly specialists*. Chicago: University of Chicago Press, 1985.

Russell, William Howard. *My diary in India, in the year 1858–9*. London: Routledge, Warne, 1860.

Said, Edward. *Orientalism*. New York: Vintage Books, 1979.

Said, Edward. *Culture and Imperialism*. New York: Alfred Knopf, 1993.

Schmidt, Heicke. 'The Notorious Emil Brugsch: "It is said that Brugsch Bey would sell the whole museum."' In Neil Cooke, ed. *Journeys Erased by Time: The rediscovered footprints of travellers in Egypt and the Near East*, 81–99. Durham and Oxford: ASTENE and Archaeopress, 2019.

Scott, Blake C. 'Revolution at the hotel: Panama and luxury travel in the age of decolonisation.' *Journal of Tourism History* 10:2 (2018): 146–64.

Secord, Anne. 'Science in the pub: Artisan botanists in early nineteenth-century Lancashire.' *History of Science* 32 (1994): 269–315.

Secord, James. *Victorian sensation: The extraordinary publication, reception, and secret authorship of 'The vestiges of the natural history of creation.'* Chicago: University of Chicago Press 2000.

Selim, Gehan. *Unfinished places: The politics of (re)making Cairo's old quarters*. London: Routledge, 2017.

Shapin, Steven. 'Lowering the tone in the history of science: A noble calling.' In *Never pure: Historical studies of science as if it was produced by people with bodies, situated in time, space, culture, and society, and struggling for credibility and authority*, 1–14. Baltimore: Johns Hopkins University Press, 2010.

Sheppard, Kathleen. 'Flinders Petrie and eugenics at UCL.' *Bulletin of the History of Archaeology* 20:1 (2010): 16–29. DOI: http://doi.org/10.5334/bha.20103

Sheppard, Kathleen. 'Margaret Alice Murray and archaeological training in the classroom: Preparing "Petrie's Pups."' In *Histories of Egyptology: Interdisciplinary measures*, ed. William Carruthers, 113–28. London: Routledge, 2014.

Sheppard, Kathleen. 'On His Majesty's Secret Service: James Henry Breasted, accidental spy.' *Journal of the Society for the Study of Egyptian Antiquities* 44 (2017–2018): 251–72.

Sheppard, Kathleen. 'Tea with King Tut at the Winter Palace Hotel.' *Journal of History and Cultures* 10 (2019): 67–88.

Sheppard, Kathleen. '"Trying desperately to make myself an Egyptologist": James Breasted's early scientific network.' In *Communities and knowledge production in archaeology*, eds Julia Roberts, Kathleen Sheppard, Ulf Hansson, Jonathan Trigg, 174–87. Manchester: Manchester University Press, 2020.

Shteir, Ann B. 'Botany in the breakfast room: Women and early nineteenth-century British plant study.' In *Uneasy careers and intimate lives: Women in science,*

1789–1979, eds Pnina Abir-Am and Dorinda Outram, 31–44. Rutgers: Rutgers University Press, 1989.

Sims, David. *Understanding Cairo: The logic of a city out of control.* Cairo: AUC Press, 2010.

Sladen, Douglas. *Queer things about Egypt.* Philadelphia: J. B. Lippincott Co., 1911.

Smith, Pamela Jane. *A 'Splendid Idiosyncrasy': Prehistory at Cambridge, 1915–50.* BAR British Series 285. Oxford: Archaeopress, 2009.

Sommers, Stephen, Lloyd Fonvielle, Kevin Jarre, Brendan Fraser, Rachel Weisz, John Hannah, Arnold Vosloo, Jonathan Hyde, and Kevin J. O'Connor. *The mummy.* Universal City: Universal, 1999.

Starkey, Paul and Janet Starkey. 'Introduction.' In *Egypt through the eyes of travellers*, eds Paul Starkey and Nadia El Kholy, vii–xvi. Durham, UK: ASTENE, 2002.

Starkey, Paul and Nadia El Kholy, eds. *Egypt through the eyes of travellers.* Durham, UK: ASTENE, 2002.

Stevenson, Alice. '"To my wife, on whose toil most of my work has depended": women on excavation.' In *Petrie Museum of Egyptian Archaeology: Characters and collections*, ed. Alice Stevenson, 102–5. London: UCL Press, 2015.

Stevenson, Alice. *Scattered finds: Archaeology, Egyptology and museums.* London: UCL Press, 2019.

Tam, Alon. 'Cairo's coffeehouses in the late nineteenth and early twentieth centuries: An urban and socio-political history.' PhD dissertation, University of Pennsylvania, 2018.

Terrall, Mary. *Catching nature in the act: Réamur and the practice of natural history in the eighteenth century.* Chicago: University of Chicago Press, 2014.

Thompson, Jason. *Wonderful things: A history of Egyptology, Vol 1: From antiquity to 1881.* Cairo: AUC Press, 2015.

Thompson, Jason. *Wonderful things: A history of Egyptology, Vol 2: The golden age: 1881–1914.* Cairo: AUC Press, 2015.

Thompson, Jason. *Wonderful things: A history of Egyptology, Vol 3: From 1914 to the twenty-first century.* Cairo: AUC Press, 2018.

Thornton, Amara. '"... a certain faculty for extricating cash": Collective sponsorship in late 19th and early 20th century British archaeology.' *Present Pasts* 5:1 (2013): 1–12.

Thornton, Amara. *Archaeologists in print: Publishing for the people.* London: UCL Press, 2018.

Tirard, H. M. and N. Tirard. *Sketches from a Nile steamer.* London: Kegan Paul, Trench, Trübner & Co., Ltd., 1891.

Trollope, Anthony. 'An unprotected female at the pyramids.' In *Tales of all countries*, 140–166. London: Chapman and Hall, 1867.

Twain, Mark. *The innocents abroad, or the new pilgrims' progress*, Vol. II. London: Harper & Brothers Publishers, 1899.

Vicinus, Martha. *Independent women: Work and community for single women, 1850–1920.* Chicago: University of Chicago Press, 1985.

Vicinus, Martha. *Intimate friends: Women who loved women, 1778–1928.* Chicago: University of Chicago Press, 2004.

Vickery, Amanda. *The gentleman's daughter: Women's lives in Georgian England.* New Haven: Yale University Press, 1998.

Walker, Archibald D. *Egypt as a Health-Resort.* London: Churchill, 1873.

Wallach, Janet. *Desert Queen: The extraordinary life of Gertrude Bell adventurer, adviser to kings, ally of Lawrence of Arabia.* London: Phoenix/Orion Books Ltd, 1997.

Waraska, Elizabeth. *Female figurines from the Mut Precinct: Context and ritual function.* Göttingen: Vandenhoeck & Ruprecht, 2009.

Weatherhead, Fran. 'Painted pavements in the Great Palace at Amarna.' *Journal of Egyptian Archaeology* 78 (1992): 179–94.

Weens, Sylvie 'Tales of antiquities at the Luxor Hotel.' *Ancient Egypt* (February/March 2019): 16–21.

Wild, Auguste. *Mixed grill in Cairo: Experiences of an international hotelier.* London: Sydenham & Co., 1954.

Wilkinson, Sir John Gardner. *Hand-book for travellers in Egypt; including descriptions of the course of the Nile to the second cataract, Alexandria, Cairo, the pyramids, and Thebes, the overland transit to India, the peninsula of Mount Sinai, the oases, &c. Being a new edition, corrected and condensed, of 'Modern Egypt and Thebes.* London: John Murray, 1847.

Wilkinson, Toby. *The Nile: Travelling downriver through Egypt's past and present.* New York: Random House, 2014.

Wilkinson, Toby and Julian Platt. *Aristocrats and archaeologists: An Edwardian journey on the Nile.* Cairo: AUC Press, 2017.

Wilson, John A. *Signs & wonders upon Pharaoh: A history of American Egyptology.* Chicago: University of Chicago Press, 1964.

Wind, Herbert Warren. 'Profiles: The House of Baedeker.' *The New Yorker* 51:31 (22 September 1975): 42–93.

Winlock, Herbert E. *Materials used at the embalming of King Tutankhamun*, Metropolitan Museum of Art Papers, No. 10. New York: The Metropolitan Museum of Art, 1941.

Winstone, H. V. F. *Howard Carter and the discovery of the tomb of Tutankhamun*, revised edition. Manchester: Barzan, 2006.

Woolley, C. Leonard and T. E. Lawrence. *The Wilderness of Zin.* London, 1914.

Wright, Arnold. *Twentieth century impressions of Egypt: Its history, people, commerce, industries, and resources.* London: Lloyd's Greater Britain Publishing Company, 1909.

Index

Printed in the USA
CPSIA information can be obtained
at www.ICGtesting.com
JSHW010723190524
63366JS00004B/16